The Struggle for Māori Fishing Rights

The Struggle for Māori Fishing Rights

TE IKA A MĀORI

BRIAN BARGH

First published in 2016 by Huia Publishers
39 Pipitea Street, PO Box 12280
Wellington, Aotearoa New Zealand
www.huia.co.nz

ISBN 978-1-77550-196-1

Copyright © Brian Bargh 2016

Cover Image: Two Māori boys net fishing, Waikato
ATL, WA-12556-G Photograph taken by Leo White.

Back Cover: Jackson Bay, Ref: 1870928.
Adobe Stock/tholi75

This book is copyright. Apart from fair dealing for the purpose of private study, research, criticism or review, as permitted under the Copyright Act, no part may be reproduced by any process without the prior permission of the publisher.

A catalogue record for this book is available from the National Library of New Zealand.

Printed in China by Everbest Printing Co.

Published with the assistance of

CONTENTS

viii	PREFACE	
1	INTRODUCTION	
13	CHAPTER 1	Fish, Fish, Everywhere: Māori fisheries
21	CHAPTER 2	Living Together: The Treaty of Waitangi and its guarantees about fish
27	CHAPTER 3	Making Waves: The Waitangi Tribunal
41	CHAPTER 4	An Ocean Windswell: Māori objections to government actions over fisheries
55	CHAPTER 5	An Incoming Tide: Government fisheries legislation in the 1980s
63	CHAPTER 6	First Bites: Recovering Māori fishing rights
79	CHAPTER 7	All Aboard the Trawler: Consolidating the gains

85 CHAPTER 8

Setting the Nets: Sealord negotiations

105 CHAPTER 9

Hauling in the Catch: The negotiations' aftermath

127 CHAPTER 10

Processing the Catch: Expanding Māori fisheries assets

157 CHAPTER 11

Something New: Who owns the seabed and foreshore?

175 CHAPTER 12

The New Millennium

191 AFTERWORD

199 APPENDIX

The Māori and English Texts of the Treaty of Waitangi

201 BIBLIOGRAPHY

205 INDEX

PREFACE

The story of Māori fisheries has not been told before. This book locates the Māori fisheries story in the context of the wider struggle by Māori against the excesses of colonisation, and the demands for the Crown to honour its undertakings as set out in the Treaty of Waitangi. It acknowledges the challenging role that iwi leaders play in Treaty grievance negotiations, often at a high personal cost, against the formidable power of the Crown. Nevertheless, the ultimate aim of both parties is to reach a solution that ensures a sustainable and positive future.

I was invited by Huia Publishers and Te Ohu Kaimoana* to document this remarkable story so that the people of this country, particularly our mokopuna, would know something of the struggle that had taken place over the previous 200 years.

I first became involved in the world of Treaty of Waitangi settlements when, as a trained environmental manager, I was working in the Department of State Coal Mines in the 1980s and heard from Tainui leaders about their claim to the coal being mined from their lands. This experience began a long-standing interest in what is commonly known as Treaty issues, working for the Crown – the Treaty Issues divisions of the Department of Māori Affairs and the Department of Justice, the Waitangi Tribunal, and iwi – Ngāti Awa and Te Arawa.

This book was written with the help of Te Ohu Kaimoana. In particular, Peter Douglas and the staff of Te Ohu Kaimoana have provided me with every assistance. I am indebted to the Waitangi Tribunal, Department of Justice, for ensuring that their records and reports are so readily available. Without that treasure trove of information, our history would be so much more inaccessible— as it was before the Tribunal began.

My thanks and gratitude go also to Sir Tipene O'Regan, Hon. Shane Jones, and Sir Doug Kidd, who agreed to interviews and provided me with information, anecdotes and other material for this story, and to Adam Gifford, who gave me access to the taped interviews he had recorded with Sir Graham Latimer and Sir Robert Mahuta.

Finally, I wish to thank Huia Publishers for offering me the opportunity to write this story and for the excellent support they have provided.

It seems fortuitous that, for various reasons, the compilation and writing of the Māori fisheries story has taken until 2015 to complete. A review of the Māori Fisheries Settlement has now been undertaken, and there are proposals to modify the role of Te Ohu Kaimoana and allow the fisheries settlement assets to be allocated to the various iwi recognised for fisheries purposes. This has already provoked further debate about Māori fisheries assets and how they should be managed in the future.

Brian Bargh
August 2015

*Note on the different names of Te Ohu Kaimoana

Following the 1989 interim settlement of Māori Treaty of Waitangi fisheries claims, the organisation set up to administer Māori fisheries assets was called the Māori Fisheries Commission. After the 1992 Crown-Māori agreement that settled Māori commercial fisheries claims, the name was changed to the Treaty of Waitangi Fisheries Commission or Te Ohu Kaimoana.

Over time the Commission has been referred to variously as Te Ohu Kai Moana (TOKM), Te Ohu Kaimoana (TOK), The Commission and the Treaty of Waitangi Fisheries Commission (TOWFC).

In order to prevent confusion I have referred to the Commission either as The Māori Fisheries Commission (prior to the 1992 agreement) and Te Ohu Kaimoana (TOK) from 1992 to the present time. In the book some quotations refer to TOK by its other titles.

INTRODUCTION

New Zealand is an island nation, and the seas around the country have always been rich in marine life. When Māori first arrived here some thousand or more years ago, they had access to a wide range of freshwater and sea fishes, marine mammals, seabirds and sea plants with which to sustain themselves. Together with the various on-land food sources, including birds and plants, the plants and mammals they brought with them to this country (sweet potato, yam, taro, rats and dogs), and their gardening technologies, Māori were able to thrive. The whole country was eventually explored and settled by Māori living together in family and tribal groupings that rigorously protected their particular territories and interests. It is estimated that the Māori population of the country when early visitors from Europe arrived could have been around 85,000.[1] The population of Māori in 2014 was about 675,000.

Māori developed a whole industry utilising freshwater, coastal and marine resources, as well as a wide variety of methods for harvesting, catching, processing and using these. Although Māori were seafarers and made voyages to and from the islands of Oceania, it is unlikely that they encountered any visitors from beyond this region; so it was probably surprising and alarming for coastal Māori to be suddenly confronted by strange looking vessels and visitors from Europe. These began with Abel Tasman (a Dutchman who had sailed from Batavia in Indonesia and landed first in Golden Bay, Nelson, in December 1642) and later James Cook (who arrived from Britain in 1769), followed by an increasing number of others.

As immigration from Europe increased in the late 1700s and early 1800s, the British reluctantly accepted responsibility for the emerging nation. A treaty was drafted, explained to a huge meeting of around 500 Māori, and signed at Waitangi on 6 February 1840 by the British government's representative, William Hobson, and many of the assembled Māori. Copies were then taken to various parts of the country, and it was eventually signed by around 500 Māori.

The Treaty of Waitangi provided for British sovereignty over New Zealand and for Māori to retain 'tino rangatiratanga' (authority) over those resources they wished. It was presented in both English and Māori, and there were three main clauses. From the time it was signed, debate has raged over which version was valid, whether the Treaty really means what it says, and what legal weight can be given to it. Some of these matters are discussed in this book, because one of the important resources seemingly and specifically reserved to Māori was fisheries.

This book tells the story of the struggle by Māori people for their right to the fisheries of this country. Prior to the signing of the Treaty of Waitangi, Māori collectively owned and controlled fisheries. It seems that from the very beginning of this new country, after the Treaty had been signed, Māori ownership of fisheries and other natural resources, as set out in the Treaty, was contentious. Yet in those first few years following its signing, there is little evidence that the Treaty had created a conundrum over fisheries. It was

Reconstruction of the signing of the Treaty of Waitangi, c. 1950

ATL: NON-ATL-0173. Marcus King

land that gave rise to immediate problems. Unlike land, there was plenty of fish for the taking and Māori did not dispute the new settlers catching fish or gathering shellfish. However, once fish started to become scarce, as occurred in a few localised places through over-fishing, the colonial government imposed restrictions to regulate the taking of fish. First came the 1866 Oyster Fisheries Act, which controlled the taking of oysters.[2] From then on, as the fishing industry began to grow and exports increased, pressure on fisheries right around New Zealand intensified and more laws and regulations were imposed. Māori were spectators in this process, which slowly and surely removed them from ownership and management of a commodity that they had owned and treasured for centuries.

The story of Māori fisheries stands on four pou (metaphorical pillars): Māori people and their seemingly infinite capacity to maintain their culture and traditions in the face of sustained opposition; the Treaty of Waitangi; the Waitangi Tribunal; and the courts of justice. Each of these pou is discussed throughout this book. The pou are firmly embedded in New Zealand society, which prides itself on being free and democratic.

The Treaty of Waitangi was the basis for governance of the country by the British Crown and is recognised today as the founding document of our country. However, there have been numerous and sustained breaches of the Treaty by the Crown that have detrimentally affected Māori. These breaches were generally ignored by successive governments for 135 years until 1975, when a serious effort began to document and redress those breaches with the establishment of the Waitangi Tribunal. The history of Māori grievances against Crown actions had been recorded but ignored by the currents of racial superiority and indifference. However, the tide of social justice, equality and fairness has prevailed against those currents. Today, Crown breaches of the Treaty continue to be redressed and compensation paid. New Zealand is probably the best in the world at addressing the grievances of its indigenous people. The relatively fair and democratic society that exists here has allowed Māori to express themselves and their claims against the Crown to be heard and acted upon. It has not been easy, but the four pou have remained firm, standing together as the supports of the Māori fisheries story. Probably the most significant strength, however, is that of the Māori people, who have never abandoned their search for justice and redress for the harm they suffered and continue to suffer as a result of Crown actions.

Historical overview

The importance of Māori people's assertion of their rights to fisheries is a feature of the more general struggle by New Zealand's indigenous people to make the best of colonisation. In a scene-setting essay written as a foreword to the German translation of *Hīkoi: Forty years of Māori protest*,[3] Dr Aroha Harris and Dr Danny Keenan recount the events that adversely impacted upon Māori and their reaction, which gave rise to avenues of reconciliation and redress that are the pou supporting this story of Māori fisheries:

Introduction

Though modern Māori protest is commonly understood to be a feature of the decades since the 1960s, Māori protest does have historical roots that reach back well into the nineteenth century and the heyday of colonialism in Aotearoa New Zealand.

Europe discovers New Zealand

Europe first sailed into the Māori world in 1642 when Dutch trader Abel Tasman sighted the Punakaiki Rocks, on the West Coast of the South Island. Tasman then sailed north and encountered the country's indigenous people, the Māori. Four of his men were killed when Māori attacked one of his vessels. Tasman did not land in New Zealand; instead, he returned to Europe with tremulous stories of his South Pacific discoveries.[4] One hundred and twenty seven years later, in 1769, Britain arrived for the first time, with naval captain James Cook claiming New Zealand as part of the extensive British realm. New Zealand was ripe for colonisation, he wrote, urging Britain to encourage the active settlement of its newest discovery. The British government however was reluctant to claim New Zealand as a formal colony; instead, it encouraged the expansion of economic development and trade, especially from nearby Australia.[5]

The earliest commercial visitors from Britain and Australia were the sealers, whalers and traders. From the 1790s, small sealing and whaling stations appeared around the unexplored New Zealand coastline, plying the foreshores and vast oceans. During the downtimes, trading enterprises also flourished. Sealers and whalers were rough and crude by nature. Many engaged in relentless and bloody hostilities against Māori, especially in the deep south. However, by and large, good relations were established between Māori and these early English commercial visitors. Many whalers took Māori names and married into Māori families, seeking permanence in New Zealand by purchasing land and establishing long-term trading ventures founded on British capital.[6]

Challenges to Māori authority

Whalers and traders were content to live among Māori as long-term visitors, learning their language, raising families and plying their trade. No attempts were made to change or challenge Māori modes of living or bases of customary authority. The earliest challenges came from the Wakefield settlers and later Wakefield politicians. After 1840, thousands of British migrants poured into New Zealand, sponsored by British entrepreneur Edward Gibbon Wakefield. Six major Wakefield townships were established in New Zealand, primarily in the North Island with settlement spreading much later to the south. The new towns were settled by pioneer migrants with high expectations of individual land ownership as the basis for establishing transplanted communities and economies. Disputed land sales with displaced local tribes however led to sharp conflicts, with aggrieved Māori facing wholesale losses of land and commensurate authority.[7] Issues such as these had first arisen in the 1830s, exacerbated by frontier lawlessness and inter-tribal violence. As a consequence of such concerns, the British Crown proposed that Māori sign a Treaty setting the terms of formal British entry into New Zealand. In return for Māori agreement to the Crown's limited role in governance, the Treaty guaranteed to Māori their customary land holdings and authority. The Treaty was signed at Waitangi on 6 February 1840, and thereafter taken around New Zealand for Māori to consider, and to sign if so persuaded.[8]

Within three years, the first violent clash over disputed land titles between settlers and Māori occurred in the South Island, at Wairau in 1843, with twenty two settlers and two Māori killed.[9] Two years later, full-scale war broke out when Māori in the far north attacked Kororareka, and especially its British flag, in 1845. This attack followed years of Māori disenchantment with the Treaty, and with their significant loss of authority in the face of determined British encroachment upon Māori sovereignty after 1840. British Army Regulars arrived in Kororareka to fight this war in 1845. Having first arrived in Wellington in 1841, the British Army fought against Māori in New Zealand for almost twenty five years, eventually withdrawing in 1865.[10]

New Zealand's wars

The wars in New Zealand, which lasted from 1843 to 1872, marked the beginnings of sustained Māori resistance against the unilateral imposition of Crown authority in New Zealand. Historians generally agree that the wars constituted a violent contest between Māori and the Crown over land and sovereignty. The wars were sporadic and affected most of the North Island, comprising nine separate but interlinked fields of engagement with the British Army engaged in seven of the campaigns. The most decisive campaign was fought in the Waikato between 1863–1864, leading to the defeat of the Māori King Tawhiao at Rangiriri which brought defeat for all Māori, though the wars would linger for a further nine years.[11] During the war period, New Zealand attained constitutional independence from Britain, in 1852, with full legislative autonomy achieved by 1856. In 1862, the government introduced the first of many Acts of Parliament designed to break up customary Māori holdings and extinguish native [land] titles, thereby breaching protections as set out in the Treaty of Waitangi. Decades of like legislation followed comprising land confiscations, land seizures for public works, suppression of rebellion and facilitation of Native Land Court activity, all aimed at depriving Māori of land and restraining Māori protest.[12]

Towards the end of the war period, in 1867, Māori were finally granted the right to vote, having waited for twenty seven years after the Treaty of Waitangi was signed. Māori were granted four special seats in the New Zealand Parliament. In 1868 the first four Māori Members were elected. Representation in Parliament changed the nature of Māori protest; with the battlefield no longer an option for Māori, Parliament became a new focus for concerted political activity.[13]

The rise of Māori political protest

The earliest Māori protest movement was the Repudiation Movement which arose in the Hawkes Bay in the early 1870s. This movement protested against unlawful land losses and sought a Commission of Enquiry into Crown

OPPOSITE PAGE

On the road to Wellington – the Māori Land March gathers pace, October 1975.

ATL: F-9-35mm-C

land purchase activities. A subsequent enquiry failed to address Māori concerns that they were increasingly the victims of iniquitous law and sharp land purchase practices.[14] More organised protests arose in the 1880s. Māori actively opposed surveys at Parihaka, in central Taranaki, and at Murimotu in Whanganui. In 1884, the Māori King Tawhiao travelled to London to present a petition to Queen Victoria. A Māori Member of Parliament, Wiremu Te Wheoro (Western Māori) resigned because the government refused to treat Tawhiao's petition seriously. New protest movements like the Orakei Parliaments in Auckland also arose. The most significant Māori protest movement of this era however was the Te Kotahitanga Parliament founded in 1888. Through Te Kotahitanga, Māori sought to establish an autonomous Māori governing body outside of, but existing alongside, the New Zealand Parliament. Supporters of Kotahitanga argued strenuously for Māori political independence, as guaranteed in the Treaty of Waitangi. By the turn of the century however, the Kotahitanga movement had tragically collapsed, riven by internal disagreements, especially after a number of senior leaders accepted further Crown guarantees over the security of land ownership and management, guarantees that were ultimately short-lived.[15]

The New Century

Land losses continued unabated after 1900, especially with Te Kotahitanga having ceased to function. Crown guarantees that Māori land ownership would be secured through new legislation were not followed through. At this time, Māori were also severely affected by falling standards of health and hygiene. The Māori census of 1896 had recorded the lowest aggregate population of Māori ever witnessed, at 40,000 Māori, with census enumerators reporting dire conditions in Māori villages.[16]

After 1900, national Māori political activity waned significantly as local Māori focussed upon improving village health standards and retaining tribal lands in tribal hands; 'Māori protest' was replaced by 'Māori survival'. Maintaining both priorities – health and lands – were sustained against great political odds and continuing iniquitous legislation, like the 1909 Native Lands Act which consolidated 40 statutes affecting Māori land and led to the resumption of systematic and wholesale land purchasing, constituting one of the largest 'land-grabs' in New Zealand history.[17]

When war broke out in Europe in 1914, a contingent of Māori volunteers was recruited and sent to Gallipoli and France as non-combatants, digging trenches, laying cables and burying the dead.[18] Other Māori however who refused to enlist were imprisoned.[19] After the war, Māori returned to their rural holdings and continued the struggle to sustain viable livelihoods on uneconomic land blocks. In the early 1920s, huge numbers of Māori were attracted

OPPOSITE PAGE

Māori Land March 1975: moving down Lambton Quay to Parliament, Wellington, 13 October 1975. Māori outrage at continuing breaches of the Treaty of Waitangi by the Crown had led to the march of protest. It was the first of numerous Māori protest actions around the country in succeeding years, including reaction against the Crown denial of fishing rights.

Fairfax NZ: Ray Pigney

to the new Ratana Movement, established by Whanganui farmer Wiremu Ratana who sought to unite all Māori, across tribal divides, in a common cause aimed at rectifying Māori impoverishment, land losses, health inequities and economic deprivation. In 1936, the Ratana Movement aligned itself to the New Zealand Labour Party and thereafter sought to win the four Māori Parliamentary seats, a feat achieved in 1943.[20] Such united Māori representation in Parliament, under the aegis of Ratana, undoubtedly advanced the political protests of Māori that had commenced a century earlier. However, at the same time, a new reality was dawning upon Māori and the country; that Māori customary land holdings had diminished in size to the point where the land could no longer support Māori livelihoods. The Crown, which had itself contributed significantly to this situation by failing to restrain land acquisitions and refusing Māori finance to develop and improve what holdings remained, now accepted that Māori must move into the urban centres. The rapid urbanisation of Māori followed over the ensuing decades. With the advent of urban living, with all of its significant social and cultural challenges for Māori, came a new basis for dramatic politicisation.[21]

Past issues of loss, dispossession and political disempowerment could now be litigated once more, within a new era and political milieu, revitalised by a new generation, themselves forced to move away from the tribal hearth because of that very same dispossession and loss. The age of modern Māori protest had begun.

Introduction

Indigenous peoples worldwide began to assert their rights, particularly in the United States of America, Canada and New Zealand, after the Second World War and particularly in the 1960s. Similarly, the remaining colonies of many countries, such as Germany, the Netherlands, Spain, France and Britain, sought independence from their overlords in that period.

Along with this increasing assertion of their rights, indigenous peoples also sought to settle long-standing claims against their governments. Aroha Harris has detailed the situation in New Zealand.[22] Under pressure from a growing Māori protest movement, which began challenging many facets of New Zealand life, the government of the day established the Waitangi Tribunal in 1975. The main function of the Tribunal is to investigate Māori claims against the Crown for breaches of the Treaty of Waitangi. The Tribunal hears evidence on these claims (initially limited to contemporary claims but later, from 1985, all Māori claims) and makes findings and recommendations to the government. The recommendations often include suggestions of redress. The Tribunal is critical to the story of Māori Fisheries.

The New Zealand courts (the High Court, Court of Appeal and, until 2004 when the New Zealand Supreme Court was established as the highest court, the Privy Council) have all played an influential and often pivotal role in adjudicating on Māori fisheries claims. However, since the Waitangi Tribunal began reporting, the Courts have often relied on Tribunal findings to assist them in coming to a decision. The Tribunal findings were critical to the final outcome of the Māori Fisheries story.

ENDNOTES

1. http://www.fourcorners.co.nz/new-zealand/new-zealand-history

2. http://fs.fish.govt.nz

3. Aroha Harris, *Hīkoi: Forty Years of Māori Protest*. Huia Publishers, 2004, translated into German as. *Hikoi: Der Lange Marsch der Maori* [Hikoi: The Long March of the Maori]. Orlanda Verlag, Berlin, 2012.

4. Anne Salmond, *Two Worlds. First Meetings Between Maori and Europeans 1642–1772*. Viking, 1991. pp. 63–86.

5. See Salmond, 1991, pp. 87–298; also Vanessa Collingridge, *Captain Cook: The Life, Death and Legacy of History's Greatest Explorer*. Ebury Press, 2002.

6. Angela Caughey, *The Interpreter: The Biography of Richard 'Dicky' Barrett*. David Bateman, 1998.

7. J. Lee, *The Old Land Claims in New Zealand*. Kerikeri: Northland Historical Publications Society, 1993.

8. Claudia Orange, *The Treaty of Waitangi*. Allen and Unwin/Port Nicholson Press, 1987.

9. Tony Simpson, *Te Riri Pakeha. The White Man's Anger*. Alister Taylor, 1979. p. 79.

10. James Belich, *The New Zealand Wars and the Victorian Interpretation of Racial Conflict*. Auckland University Press, 1986. p. 21.

11. Tom Roa, 'Kingitanga', in Malcolm Mulholland and Veronica Tawhai (eds.), *Weeping Waters. The Treaty and Constitutional Change*. Huia Publishers, 2010. pp. 165–173.

12. Alan Ward, *A Show of Justice. 'Racial Amalgamation' in Nineteenth Century New Zealand.* Auckland University Press, 1995 (Reprinted with corrections).

13. Keith Sinclair, *Kinds of Peace. Maori People After the Wars, 1870–8.* Auckland University Press, 1991. pp. 86–98.

14. Sinclair, 1991, pp. 112–120.

15. Alan Ward, *An Unsettled History. Treaty Claims In New Zealand Today.* Bridget Williams Books, 1999. p.147; Basil Keane, 'Kotahitanga', in Maria Bargh (ed), *Maori and Parliament: Diverse Strategies and Compromises.* Huia Publishers, 2010. pp. 9–16.

16. R. E. Brown, Registrar-General, Maori Population, *Appendices to the Journals of the House of Representatives,* 1875, G-2, p. 20; R. Parris, Memorandum to Under Secretary, Native Department, *Appendices to the Journals of the House of Representatives*, 1881, G-3, p.12; R. W. Woon, Resident Magistrate's Report, *Appendices to the Journals of the House of Representatives,* 1875, G-2, No. 13, p. 14.

17. Ward, 2010, pp. 156–158.

18. Rikihana Carkeek, *Home Little Maori Home, A Memoir of the Maori Contingent 1914–1916.* Totika Publications, 2003; Christopher Pugsley, *Te Hokowhitu a Tu: The Maori Battalion in the First World War.* Random House, 2006; E. Keenan, 'A Maori Battalion: The Pioneer Battalion, Leisure and Identity 1914–1918', unpublished BA (Hons) thesis in history, Victoria University of Wellington, 2007.

19. Michael King, *Te Puea Herangi: From Darkness to Light.* Department of Education, 1984. p. 25.

20. Ranginui Walker, *Ka Whawhai Tonu Matou Struggles Without End.* Penguin Books, 1990. pp. 195–196.

21. Walker, 1990, pp. 197–199.

22. Harris, 2004.

FISH, FISH, EVERYWHERE:
Māori fisheries

In this chapter the 'Māori fisheries' referred to in the Treaty of Waitangi are described. While it is clear that Māori tribes had authority over (that is, owned) all of New Zealand prior to the signing of the Treaty, through that signing and ratification the tribes entered into an arrangement that has been much discussed, analysed and defined ever since. Two major questions are discussed here: what were Māori fisheries, and how did Māori lose control of their fisheries, having been guaranteed the 'exclusive and undisturbed possession' of them by the Treaty of Waitangi in 1840?

While Article Two of the Treaty guaranteed Māori their rangatiratanga over 'nga taonga katoa' (valued possessions, including fisheries), Article One of the Treaty allowed for the Crown to make laws, including laws about fisheries. As with other natural resources (such as land, forests, water and waterways), differences of opinion arose between the Crown and Māori over interpretation of the Treaty's wording. The extent to which either Treaty partner was entitled to 'control' fish and fishing (and other natural resources) was disputed.

New Zealand waters, both fresh and salt, abound with many different fish species, and huge value was placed upon them at the time of the Treaty. In the Manukau, Muriwhenua Fishing and Ngai Tahu Sea Fisheries reports, the Waitangi Tribunal was presented with copious volumes of evidence, both oral and written, as to the physical (species, extent), legal, commercial, customary and traditional nature and use by Māori of their fisheries. In their reports, the Tribunal addressed each of these aspects in 'findings' which are an invaluable guide to understanding the situation today.

The term 'Māori fisheries' includes not only physical and social components, but also a legal description derived from the Treaty of Waitangi. However, once the Treaty was signed and new settlers began arriving in great numbers, Māori rapidly became a dwindling minority in their own country. The words of Article Two, guaranteeing Māori 'undisturbed possession' of all their 'properties' (English language version), and 'tino rangatiratanga' (full authority) over 'nga taonga katoa' (Māori version), including fisheries, soon became null. In its Kaituna River Report, the Waitangi Tribunal quoted from evidence given by PG McHugh:

> By 1840 it was a settled principle of colonial law that the land rights of aboriginal people were protected by the Crown as evidenced by Lord Normanby's Instructions to Hobson: ". . . (the Maori) title to their soil and to the sovereignty of New Zealand is indisputable and has been solemnly recognised by the British Government . . ." For nearly forty years after the signing of the Treaty, New Zealand Courts recognised and accepted these principles of colonial law as to the land rights of the Maoris.[1]

However, in 1877 Chief Justice James Prendergast ruled that, effectively, unless the Treaty was enshrined in legislation it could not be enforced. The judgement[2] involved Māori land in Porirua (near Wellington) that had been given to the Anglican Church for the purpose of building a school. The school was never built and Wiremu Te Kakakura (Wi) Parata asked that the land be returned to his tribe, Ngāti Toa. Octavius Hadfield, the Anglican Bishop of Wellington, opposed the request and the case went to the Supreme Court. Prendergast took the view that 'native' or 'aboriginal' customary title, not pursuant to a Crown grant, could not be recognised or enforced by the courts, because the Treaty of Waitangi was a 'simple nullity'. He described Māori as 'primitive barbarians', and said they were 'incapable of performing the duties, and therefore of assuming the rights, of a civilised community'.[3] Later, when fisheries issues were being dealt with by the Crown and Māori challenges were being made, the Crown interpretations based on that Supreme Court judgement prevailed, and have done so until very recently.

Pre-settler Māori fishing

The term 'Māori fisheries' has been outlined by the Waitangi Tribunal in a number of reports. One of the first Tribunal hearings, in 1983–4, was about a proposed sewage outfall into the Kaituna river near Rotorua. The Tribunal described the Kaituna river as having been owned for many generations by Ngāti Pikiao, whose traditional rights included the right to fish the river, the estuary and the sea, free and uninterrupted. It stated that these rights 'continued uninterrupted to this day', and it added that the Treaty of Waitangi guaranteed the continued enjoyment and undisturbed possession of 'Taonga Maori' (treasured possessions), of which fisheries was one.[4] In the Manukau Report, the Tribunal found that in the Manukau Harbour, 'Maori fisheries are extensive and indeed, the whole of the Manukau could be described as a traditional Maori fishery . . . '.[5] Similarly, the Tribunal found in the Motunui-Waitara Report that hapū of Te Āti Awa had 'significant and traditional fishing grounds', and that their use of these had continued unbroken into modern times.[6] However, the most detailed accounts of Māori fishing and fisheries are in the Muriwhenua Fishing Report released in May 1988 and the Ngai Tahu Sea Fisheries report released in August 1992.

The Muriwhenua and Ngāi Tahu fisheries

Muriwhenua is that area of land in the Far North of New Zealand. The tribes there claimed that the Crown had breached their rights to their fisheries as guaranteed by the Treaty of Waitangi. How was it, the Tribunal asked in its report, that:

> . . . a people who once depended so heavily on the sea resource should now [in 1987] find themselves almost totally shut out of economic activity which was so much a part of their way of life . . .[7]

> . . . The Muriwhenua tribes made a full and extensive fishing use of the fresh waters and seas within and around their mainlands and islands, including the swamps, lakes, rivers, inlets, estuaries, harbours, foreshores, beaches and seas. That use is evidenced by the substantial knowledge of fishing lore, grounds and methods retained by Muriwhenua people of today and by various written accounts.

> An intensive all year fishing use was made of the seas to about three miles off-shore. This band includes the bulk of the modern rock lobster fishery, all of the paua, kina, other shell-fish and the seaweed resources, and large proportions of many of the prime fin-fish fisheries, the Northland long-line snapper fishery for example, and all set-net and seine fisheries for species like mullet, flounder, rig, school shark, moki and kahawai.

> Throughout the balance of the continental shelf, to about 12 miles from the shore, fishing was intensive and regular but mainly seasonal. Expeditions coincided with the offshore migrations of such species as hapuku, bass and snapper. Also fished were species more typical of off-shore areas such as tuna, pelagic sharks, tarakihi, piper, mackerel and squid. The Muriwhenua continental shelf, as fixed by a 200 m depth contour, closely approximates New Zealand's 12 mile territorial sea. The Muriwhenua people generally fished in depths not exceeding 200 m.[8]

LEFT

Great moments in New Zealand history no 1. Well done Hobson. With a bit of luck we'll never hear about fishing rights or land claims ever again.

Cartoon. Shows symbolic walking away from the table (representing unity) by the Crown (one way) and Māori in the opposite direction. Cartoon published at a time the Waitangi Tribunal had released the controversial Muriwhenua Fishing Report in 1988.

ATL: H-733-117. Thomas Scott, Evening Post, *28 September 1988.*

BELOW

Unidentified woman probably gathering oysters at Paihia, c. 1900. Māori oyster beds were the first 'fish' to be brought under government control in the 1860s in order to conserve them.

ATL: G-8468-1/2. Frank James Denton

The Ngāi Tahu fishery includes the lagoons, harbours, estuaries, rivers and streams of their rohe (territory) and the sea off their coastline, which extends around most of the South Island, except for the northern area. The Tribunal report on the Ngāi Tahu claim states that:

> Fishing formed an essential part of the Ngai Tahu economy prior to the Treaty, as did the taking of seals, shellfish, whales and marine flora. The use of marine resources was a fundamental feature of Ngai Tahu mahinga kai (places where food is), and played a key role not only in the tribe's economy, but in its social and spiritual life. Marine resources formed a significant part of the diet of Ngai Tahu communities . . .[9]
>
> . . . Ngai Tahu also made considerable use of the fisheries in their dominion, particularly in certain spots up to a mile from the shore and less frequently out to about twelve miles from the shore.[10]

Traditional fish management

The Tribunal reports also outline traditional fish management. The Muriwhenua Report states that:

> Rights of passage in the inner fishing band were restricted to the hapu and tribes of Muriwhenua and those having peaceful business with them; and that restriction was enforced or was capable of being enforced. Rights over fishing and passage were claimed in respect of the outer band but may have been recognised by others only to the extent that they could in practice be enforced. In the Maori idiom the hapu and tribes of Muriwhenua held the 'mana' or 'authority' of the whole of the Muriwhenua seas to the limit described. The nearest British cultural equivalent is to consider that they exercised 'dominion' over that part, or 'owned' it as part of their territorial waters.[11]

Similarly, the Tribunal found Ngāi Tahu had:

> . . . developed a sophisticated fishing technology that was adapted to the various marine environments of Te Waipounamu (the South Island) and to the particular species to be found in them. The development of barracouda lures and composite hooks, as well as the use of a variety of small nets, allowed Ngai Tahu to take species not readily available to North Island Maori.[12]

The Tribunal comments that 'the evidence is rather compelling that Maori in many places throughout the New Zealand islands worked the whole of the inshore seas'.[13] It stands to reason that although various hapū may have had different tikanga (protocols) about their fisheries, the inland and coastal waters of New Zealand were plentiful in fish and sea-plant life, and there is no reason to suppose that hapū would have widely differing catch methods, usage or values placed on their fishery. Indeed, this point is covered by the Tribunal in making findings as to the value of the Muriwhenua fisheries and their extrapolation to most of the rest of New Zealand, up to and around the time of the Treaty:

> The common cultural characteristic of the Maori tribes was the paramount dependence upon the products of an aquatic economy. Their fisheries had subsistence, commercial, recreational and cultural aspects. They were essential not only for physical survival of individuals and communities but the whole economies and social networks of the hapu and tribes involved. The commercial component that existed in pre-European times was capable of adaptation to commercial uses in Western terms. The sea resource was essential for the physical survival of the Maori people and their communities. It supplied almost the only animal food they could obtain, providing essential proteins, fats, vitamins and minerals. A fishing ground could be of much greater value and importance to their existence than any equal portion of land. Fishing was important for all tribes, but the lack of comparable inland resources in Muriwhenua made the sea resource more important for them than for most others. Their dependence on the sea was greater. The sea resource was as important for their survival as the atmosphere they breathed.[14]

The development potential of traditional fisheries

The Tribunal was clear that:

> There was a commercial component in pre European tribal fisheries through 'gift exchange'. Though conducted along distinctive lines it was trade and commerce nonetheless. Artefacts show that the Muriwhenua tribes traded widely. Gift exchange was capable of adaptation to new circumstances. It in fact adapted and developed to trade in Western terms. There are no customary constraints on the development of a trade in fish save those cultural inhibitions directed to resource maintenance. There are no traditional constraints on the use of new technologies. There is no rule that those practices handed down must be passed on without improvements . . . Subject to the inhibitions described, there were no customary constraints to prevent the development of fisheries in Western terms once settlement came.[15]

Indeed, the Muriwhenua Māori stepped up fishing to sell the catch to the new settlers. However, fishing in the Far North was primarily for survival, as there was a limited land area for subsistence. Similarly, after European contact Ngāi Tahu were able to:

> . . . adopt technological advances suited to their fishery such as iron barbed hooks, oars and rollocks and eventually sealing and whale boats. This process was well under way prior to the Treaty, and continued in the decades immediately following. Ngai Tahu traded fish amongst themselves and with other iwi and they preserved fish for their own consumption and for trade. With the coming of Europeans they had the opportunity to trade fish with visiting vessels and to supply European whaling shore stations. Some Ngai Tahu fish may have been exported at least as far as Australia by the late 1830s.[16]

The Tribunal was also clear that Māori did not object to non-Māori use of the seas, provided that this did not interfere with Māori fishing. In fact Māori considered themselves as retaining their authority over their seas – as the Treaty had guaranteed they could. 'Other than whaling, there was [at that time] no evidence of any non-Maori commercial fishing'.[17]

At 1840, therefore, Māori were guaranteed (by the Treaty), and had exercised 'effective tino rangatiratanga' or 'exclusive and undisturbed possession of their fisheries'. 'Their fisheries' were defined by the Tribunal as 'their activity and business of fishing', including 'the fish that they caught, the places where they caught them, and the right to fish. They were not limited to site specific grounds, favourite fishing places or a mere right of access to the sea.'[18]

When the new European settlers began to arrive, Māori were able to include them in their trading operations. Rather than bartering among themselves and with other tribes, a cash economy immediately developed. The Tribunal commented:

> For some 25 years post-Treaty, fishing was not an issue. The settlers were concerned with land. Maori were unrestricted in their fishing and fish trade and they in turn had no reason to seek limits on the settlers' fishing, for the settlers fished mainly for their subsistence and personal needs. Then, somewhere in the historical process, the roles became reversed.[19]

The Muriwhenua Report identified the 1860s as 'marking the turn of the tide':

> It was then that the numerical superiority of the settlers was achieved. It was also at that time that Britain passed over to them its political control, and war with certain Maori was declared. Racial attitudes hardened. In the wake of the wars came a series of laws destined to break the Maori control of the resources of the land and sea, and significantly, to put an end to their competitive trading habits.

In the area of fishing those laws related first to oysters. The Oyster Fisheries Act of 1866 was targeted at the supply of oysters to Auckland.

Less than one year beforehand, the House of Representatives had been furnished with a return showing that Maori had supplied to the settlement literally thousands of kits of oysters. Government forbad the commercial exploitation of oysters by Maori, and leased Maori oyster beds to non-Maori commercial interests. By subsequent Acts, Maori would be protected, it was considered, for provision was made for Maori oyster reserves. But none was reserved, at least not before 1913 and only after the local beds had been severely depleted by non-Maori pickings.

The more significant feature was that Maori were prohibited from selling oysters from beds reserved for them. Those beds were for personal needs alone, for that was what tradition was said to imply. So was the view first established that the Maori interest in fisheries was non-commercial, and could be provided for by the reservation of a few fishing grounds. The assumptions made at that time permeated subsequent fishing laws as inland fisheries (from 1867) and then marine fisheries and fisheries as a whole (from 1877) were brought within the purview of statutory regulation. The regime was continued in all fishing laws, thereafter, to the present day.[20]

The Ngāi Tahu report similarly makes it clear that in Te Waipounamu, until the 1860s, Ngāi Tahu 'continued fishing without any significant involvement by Europeans apart from diminishing activity by Europeans involved in sealing and whaling':

During the 1840s and 1850s as settlers arrived in Otago and Canterbury in increasing numbers Ngai Tahu actively traded with them in the supply of sea fish. Ngai Tahu also continued to trade on a gift-exchange basis among themselves. As European settlement built up in the 1850s and 1860s a viable market for a Ngai Tahu commercial fishery developed.

By the mid 1860s . . . Ngai Tahu commercial fishing extended out as far as 20 to 30 miles from the shore in some locations with the aid of marks books [fishing marks were based on marks books recording land marks which, when aligned, allowed fishers to relocate specific fishing sites with considerable accuracy]. Whaleboats or adaptations of these were used, often fitted with sails. These came to be favoured by European commercial fishermen also. They were the basis of the South Island fishery until early this century. European commercial fishing began slowly in the 1860s . . . [and] accelerated in the 1870s and continued to grow thereafter. However, by the 1880s some over-fishing became apparent (for example in the Otago and Akaroa harbours and certain oyster beds) . . .

. . . Following the catastrophic effects of the Crown acquisition of virtually all their land, commercial fishing came increasingly to be carried on by individual Ngai Tahu. They fished commercially for their livelihood either singly or in association with family members and with the aid of closely guarded marks books. The individual Ngai Tahu who were so engaged were often descended from European whalers and had access to whale boats and associated fishing equipment, whereas following their post-land sales impoverishment, most Ngai Tahu whanau or hapu lacked the necessary capital to engage in commercial fishing.[21]

Into rougher waters

At the beginning of this chapter two questions were asked: what were Māori fisheries, and how did Māori lose control of their fisheries, having been guaranteed the 'exclusive and undisturbed possession' of them by the Treaty of Waitangi in 1840? Drawing largely on evidence presented by Muriwhenua, Ngāi Tahu, Manukau and other Māori in a series of Waitangi Tribunal inquiries, and

OPPOSITE PAGE

Wharerau Whaitiri hanging out green eels to dry at Te Roto o Wairewa (Lake Forsyth), Canterbury, 1948.

ATL: F-40048-1/2. KV Bigwood

the Tribunal findings based on that evidence, the answers are clear. They apply to all coastal Māori and Māori with access to coastal areas.

Fish and other sea-life were of paramount value to Māori, being essential for life. Over time, fishing technologies that Māori brought to New Zealand were adapted to capture and use a large range of fish, shellfish and other marine and freshwater life. In the years after 1840, non-Māori were able to utilise these resources for themselves. As the 'cash economy' in fish developed, by about 1870 some local fisheries were showing signs of overfishing. This brought some species to the attention of the government, which assumed control and imposed laws, ostensibly for the conserving and protection of threatened fisheries. The right of the government to make such laws without Māori agreement and the subsequent reaction by Māori are the subject of the next chapter.

ENDNOTES

1. Waitangi Tribunal, *Kaituna River Report*. Department of Justice, 1984. p. 31.

2. *Wi Parata v the Bishop of Wellington* (1877) 3 NZ Jur (NS) SC 72.

3. http://en.wikipedia.org/wiki/Wi_Parata_v_Bishop_of_Wellington.

4. Waitangi Tribunal, *Kaituna River Report*. Department of Justice, 1984. p. 31.

5. Waitangi Tribunal, *Manukau Report*. Department of Justice, 1983. p. 83.

6. Waitangi Tribunal, *Motunui-Waitara Report*. Department of Justice, 1983. p. 53.

7. Waitangi Tribunal, *Muriwhenua Fishing Report*. Department of Justice, 1988. p. xi.

8. Waitangi Tribunal, 1988. pp. 196–97.

9. Waitangi Tribunal, *Ngai Tahu Sea Fisheries Report*. Department of Justice, 1992. p. 289.

10. Waitangi Tribunal, 1992. p. 290.

11. Waitangi Tribunal, 1988. p. 198.

12. Waitangi Tribunal, 1992. p. 289.

13. Waitangi Tribunal, 1988. p. 197.

14. Ibid, p. 200.

15. Ibid.

16. Waitangi Tribunal, 1992. p. 291.

17. Waitangi Tribunal, 1988. p. 201.

18. Waitangi Tribunal, 1992. p. 292.

19. Waitangi Tribunal, 1988. p. xv.

20. Ibid, pp. xv–xvi.

21. Waitangi Tribunal, 1992. p. 293.

LIVING TOGETHER:
The Treaty of Waitangi and its guarantees about fish

Since the introduction of the first fisheries laws, Māori have challenged Crown actions and legislation in the courts through direct action, petitions to Parliament, meetings with Crown agents and claims to the Waitangi Tribunal. In the various judgements and findings, certain fundamental principles have been articulated, some peculiar to the specific matter being argued and some in the form of more general, widely applicable principles.

As noted earlier, the Treaty is one of the important foundations (pou) of this story, but the wording of the Treaty itself has been the cause of much debate. The English text is not an exact translation of the Māori text [see Appendix I for the two texts]. Despite the problems caused by the different versions, both represent an agreement in which Māori gave the Crown the right to govern and to develop British settlement, while the Crown guaranteed Māori full protection of their interests and status, and full citizenship rights.

This chapter draws on deliberations by the Waitangi Tribunal and the New Zealand Law Commission. Despite the large volume of academic and not-so-academic work done on the Treaty concerning, for example, its interpretation, whether it was an expression of indigenous rights and so on, a working model of how the Treaty should operate can be drawn up from these deliberations. This chapter is not a dissertation on the Treaty itself, but rather a distillation of information and legal opinion as it applies to the story of Māori fisheries. It is in the nature of a fishing expedition.

The Treaty of Waitangi and fisheries

When the British decided to enter into an agreement with Māori, they were faced with some facts: Māori tribes owned and occupied all of New Zealand – there was no 'wasteland'. British colonial policy in the 1840s was based on humanitarian principles; Māori were numerous and powerful. This led the British to present a treaty to Māori because this was both an honourable and a pragmatic pathway to colonisation.[1] However, as the Law Commission has noted, 'what was accepted for land was not applied to fishing grounds below the high tide mark'.[2] Despite the Treaty of Waitangi seeming to grant ownership to Māori of their fisheries for as long as they wished, the Crown's stance was that common law rules applied and, if necessary, overrode the Treaty.

One of these common law rules was that land below the high tide mark belonged to the Crown. There was therefore, said the Commission, 'a general right to fish both the tidal and offshore waters. These rules were taken by the courts and government to prevail over any Treaty promises or principle of aboriginal title . . . to many Maori it was novel and alien. It lies at the heart of many Maori grievances over fishing rights.'[3] The Commission went on to outline a fundamental issue concerning fisheries:

> So, in the past, Maori interests in fisheries have been seen as essentially personal and social, pertaining to subsistence and to hospitality. This has been the assumption of governmental and administrative policies and regulatory measures.

Special reserves and exemptions from general regimes have been directed at allowing use for personal consumption or at hakari, hui and tangi. These uses are important. But the implication that there was no wider or more general Maori interest in fishing is inconsistent with a great deal of evidence.[4]

The Waitangi Tribunal is specifically responsible (Treaty of Waitangi Act 1975) for determining the meaning of the Treaty, taking into account the Māori and English versions. The Tribunal has found that the Treaty:

> ... was a valid treaty under international law. Certainly it was the intention of the British government to treat with the Maori people as a sovereign independent nation. Accordingly it is reasonable to apply the general principles of treaty interpretation to the Treaty of Waitangi ... the broad and general nature of its language indicates the Treaty was not intended as a finite contract but rather a blueprint for the future. For all lay in the future. What matters is the spirit. It is necessary to look not only at the language of both texts of the Treaty but also to the surrounding circumstances including the Maori perception at the time of what the Treaty meant.[5]

The 1989 Law Commission Report, *The Treaty of Waitangi and Maori Fisheries,* also took up this point:

> ... it is the essence and spirit of the Treaty that should be looked at in considering the Crown's obligations ... The Treaty gave the Crown what it sought: sovereignty and governance over New Zealand. What Maori received in return is likewise ongoing — the continued protection of the rights that the Treaty acknowledged as theirs. Among those rights were the fisheries of Maori tribes.[6]

The Waitangi Tribunal found that:

> With very few exceptions the Maori version of the Treaty was signed by the Maori chiefs. Where there is a difference between the two versions considerable weight should, in our opinion, be given to the Maori text since this is the version assented to by all but a few Maori who signed the treaty.

> We believe the Treaty of Waitangi should be seen as a basic constitutional document ... and ... must be capable of adaptation to new and changing circumstances as they arise.

> The High Court has ruled that the Treaty 'is part of the fabric of New Zealand society' and in certain circumstances regard may be had to its provisions in interpreting legislation. But in the absence of express legislative provision, Treaty rights cannot be enforced in the courts.[7]

In its Ngai Tahu Fisheries Report, the Tribunal declared that:

> The cession by Maori of sovereignty to the Crown was in exchange for the protection by the Crown of Maori rangatiratanga. This principle is fundamental to the compact or accord embodied in the Treaty and is of paramount importance. We see it as over-arching and far-reaching.[8]

> It obliged the Crown to protect Maori rangatiratanga over their rights and properties (lands, forests and fisheries in particular) in return for the right, granted by Maori, to govern the country in the interests of all New Zealanders. Inherent in this principle to 'protect' was the obligation on the Crown to ensure their protection extended to Maori rights to develop their fisheries (and other resources) into the future. In other words, these rights Maori had were not static and did not apply to things as they were in 1840. 'Protection' meant the Crown 'had an active duty to protect to the fullest extent possible.'[9] The Crown in the exercise of its powers of governance in the national interest clearly has a right, if not a duty, to make laws for the conservation and protection of valuable resources such as the sea fisheries. But such power should be exercised with due regard to the interests of the owners of such resources. In the case of their sea fisheries guaranteed to Māori by the Treaty,

the Crown should first consult with Māori on proposed conservation measures and ensure that Māori interests are not adversely affected, except to the extent necessary to conserve or protect the resource.[10]

Perhaps the Tribunal is hinting that sea fisheries should have been shared from the outset so that both the Crown and Māori benefitted as time progressed: 'Their fisheries in the Treaty we find, refers to their activity and business of fishing, and that must necessarily include the fish that they caught, the places where they caught them, and the right to fish'.[11] So to quote the Tribunal:

> In 1840, the fish they caught in Muriwhenua, were the whole of the inshore and migratory species. The place where they caught them was the whole of the inshore seas. The business they had in fishing was extensive and capable of being developed.[12]

The Tribunal also pointed to two further principles distilled from the words of the Treaty of Waitangi that are essential to this fisheries story. The principle of partnership is founded on the idea that two peoples live in New Zealand because the Treaty 'extinguished Māori sovereignty and established that of the Crown. In so doing it substituted a charter, or a covenant in Māori eyes, for a continuing relationship between the Crown and Māori people, based upon their pledges to one another. It is this that lays the foundation for the concept of a partnership.'[13] The principle of mutual benefit arises from the fact that, 'Both parties expected to gain from the Treaty, the Maori from new technologies and markets, non-Maori from the acquisition of settlement rights and both from the cession of sovereignty to a supervisory state power.'[14]

The discussion has so far looked at the Treaty of Waitangi and winnowed out the essential arguments that explain what was intended by both Māori and the British Crown at the time of its signing, and how the different understandings of what was being agreed would take shape in the future to become matters of dispute. This happened once fish in particular localities became scarce (about 1865) and the government laid out a law to deal with over-fishing (noted in Chapter 1 and discussed further in Chapter 4).

The Law Commission has put forward an alternative way of dealing with Māori rights to fisheries. In summary, they explain that British common law applied to British territories however they were acquired (by force or cession, for example). The common law recognised land and other rights of native peoples. The Treaty is no more than a reaffirmation of Māori property rights that exist despite the Treaty. This is the concept of aboriginal title mentioned already. 'Maori property rights continued to exist unless, and until, legislation took them away.'[15] However, in its Muriwhenua Fisheries report, the Tribunal rejected the notion that aboriginal title upheld Māori fishing rights or that the Treaty of Waitangi was based upon that doctrine.[16] The Tribunal noted that the doctrine had been recognised in New Zealand in 1847 and again in 1987. The latter reference is to the case of Tom Te Weehi, which was important in the fisheries story both for what the judge determined and for its timing.

The common law assumption referred to above has an interesting history. Professor Peter Pearse (one of the architects of New Zealand's Quota Management System, discussed in Chapter 7) outlined this briefly in a paper he gave in 2007 in Iceland. He began by describing early fishing in Britain:

> From ancient times and through the early Middle Ages, much of the marine fishing in Britain and Europe actually took place in rivers and estuaries, involving weirs, traps and other fixed gear attached to stream banks and beaches. Consistent with this link to the land, rights to fisheries were held by the owners of the bordering land. A major turning point for countries that inherited British traditions of fishing and maritime law was the signing of the Magna Carta by King John of England in 1215. Landowners

had become upset when the king ignored their right to the fisheries in certain rivers and began granting fishing rights to outsiders. So the barons at Runnymede [a meadow beside the Thames river where the Magna Carta was signed] inserted a clause in the Magna Carta which effectively prevented the king from granting exclusive fishing rights in certain English tidal rivers.

Over the following couple of centuries, the courts gradually transformed the king's undertaking into a general law forbidding the king or anyone else from granting exclusive fishing rights to anyone in any tidal waters. Thereafter no one could hold exclusive rights to fish or exclude anyone else from fishing. This became known as the general 'public right of fishery' in tidal waters. This ban on private property in tidal fisheries was reinforced by the ancient 'law of capture' which ruled that no one could claim ownership over wild animals or fish until they were caught. Four centuries after King John, further reinforcement came with the doctrine of the 'freedom of the seas' articulated by the Dutch jurist Hugo Grotius in 1609; this held that no one, and no nation, could own the high sea or restrict anyone from fishing on it. In fact, unrestricted access to fisheries became a deeply entrenched principle among fishing nations, and remained the general rule until recently.

From an economic viewpoint, the open-access regime was natural and appropriate as long as the fish stocks could easily satisfy all the demands on them, as was usually the case historically. There was not a need to regulate fisheries. The open access policy was supported by the generally accepted view that ocean fisheries were inexhaustible.

But as the world population expanded and fishing fleets grew and employed new technologies for catching fish there was a need to control fishing throughout the world. So, in the twentieth century, governments of fishing nations launched a massive regulatory effort to protect stocks from overfishing. These efforts did not change the fundamental rights of fishermen, but government regulators progressively introduced closed fishing seasons, closed areas, and myriad restrictions on boats and fishing gear in an attempt to constrain catches to the sustainable productive capacity of the stocks. But given continued expansion of fishing fleets and advances in technology, these efforts had mixed success. After World War II it was clear that many of the world's most valuable stocks were fully exploited, and many were being depleted.[17]

The Law Commission clearly sets out the issue of the common law in relation to the Treaty. Despite the clear wording of the Treaty:

> the generally accepted opinion from the 1870s was that its legal effect was nil and the very acquisition of British sovereignty prevented the courts from taking account of any so-called Maori 'rights' because the Treaty of Waitangi was deemed to not be a treaty according to international law [this is a technical legal argument]. The common law was therefore said to apply and so 'all persons' had the right to fish in tidal waters, estuaries and the sea, over-riding or negating any sort of Maori ownership and control or exclusive rights to fish. Put more simply, the Treaty of Waitangi 'had no direct effect on the law of New Zealand, and did not give rise to rights that could be recognised in a court of justice.'[18]

It is clear that for many years the Crown did virtually nothing about this situation, despite the increasingly vocal calls by Māori — in Parliament, in the courts and on marae — to uphold their fishing rights, as they understood them, under the Treaty.

When sea fisheries were first made subject to statutory regulation (Fish Protection Act 1877), Māori rights were recognised in Section 8:

> Nothing in this Act contained shall be deemed to repeal, alter, or affect any of the provisions of the Treaty of Waitangi, or take away, annul, or abridge any of the rights of the aboriginal natives to any fishery secured to them thereunder.

Although Māori formally and informally challenged the government's control of fisheries in petitions, defiance of the law and in various other ways over many years, they were almost always unsuccessful.

The basic 'protection' in the law was subsequently left largely intact until the Crown introduced the Quota Management System in 1986 by amending the then Fisheries Act. This rudimentary 'protection' became a crude 'hook' that is crucial to this story.

The evidence gives rise to a vital question. Article Two of the Treaty requires protection of Māori collective rights and properties, yet those properties (fisheries amongst them) were not given protection at all except, as discussed in Chapter 4, through clauses similar to Section 8 included in Fisheries legislation as noted above. Why was this? Sir Tipene O'Regan believes that Pākehā New Zealander culture has 'always hated the tribe'. From the start of colonisation here:

> ... the tribal collective has historically been the economic competitor of the settler, the barrier to State land acquisition and, more lately, the constraint on the absolutist aspirations of the State in resource control. The Article II Treaty [of Waitangi] rights confirm Maori collective property rights. The answer is simple. Destroy the tribe and you free up the assets.[19]

O'Regan quotes Prime Minister Henry Sewell in the 1870s calling for public policy that will stamp out 'the beastly communism of the Maori'. Underlying this is the aim of denying or subverting the 'collective property right of the tribe'.[20]

In its assessment of the situation in 1986, the Law Commission took a careful look at the meaning of the Treaty of Waitangi, and particularly understandings that had been made by the Court of Appeal in its decision in the case of *New Zealand Maori Council and Latimer v Attorney General and Others*.[21] The Māori Council had gone to court seeking to prevent the then Lange Labour Government from disposing of Crown lands as part of their desire to sell various government owned businesses, such as forestry, railways and other enterprises. The Māori Council feared that in selling such lands, nothing would be left to redress historical grievances that often involved those very lands. However, not only land rights were at issue. The Court of Appeal said that the Treaty was to be applied in accordance with its spirit and intent in the circumstances of today, and not those of 1840 when it was signed. In other words, the Treaty is 'always speaking'.

Sailing forth together

The Law Commission was clear that the Treaty of Waitangi marked:

> ... the beginning of constitutional government in New Zealand. It was the means [by which] British authority and government came to New Zealand peaceably and with the consent of those this country belonged to. By the Treaty the Maori gave to the British Crown the right to make laws and govern in return for the promise to recognise and protect those things that the Maori valued ... The case for the recognition of Maori fisheries rests in large degree on respect for property rights ...
>
> ... The argument for the recognition of Maori fisheries is this. These fisheries were historically vested in iwi and hapu ... (and were a major economic resource). A condition of the consent of the chiefs to British sovereignty was that possession of these fisheries should be guaranteed 'as long as it is their wish and desire to retain' them. Fishing rights are to be respected and protected not as a privilege for Maori, but because these rights belonged to the various communities which formed the people of Aotearoa before the European came to its shores and have never been sold or given away.[22]

The Waitangi Tribunal is another of the pou (sacred pillars) upon which the Māori fisheries settlement sits. Because the Tribunal is charged with determining the meaning of the Treaty in a variety of circumstances, this chapter ends with a quote from their Muriwhenua Fisheries Report (written in 1987 and released in May 1988):

> A literal interpretation of the Treaty's terms undoubtedly creates an awkward result today; but there is and always was room for another arrangement to be settled upon. The Treaty provided for it. Nothing restricted the negotiation of alternative fishing arrangements. The current inconvenience arises not from the Treaty's terms, but from the Crown's past failure to seek or provide for a reasonable settlement. Instead, Maori fishing rights were simply denied. The Crown cannot now profit from the inconvenience that arises from its own wrong.
>
> Were the Crown to exercise its sovereign power so as to expropriate land without compensation, it would be a gross breach of the terms of the Treaty. Fisheries are in no different position.[23]

ENDNOTES

1. Law Commission, *The Treaty of Waitangi and Maori Fisheries*. Prelim. Paper No. 9, 1989. p. 23.
2. Ibid.
3. Ibid.
4. Law Commission, 1989, p 26.
5. Waitangi Tribunal, *Ngai Tahu Sea Fisheries Report*. Department of Justice, 1992. pp. 267–68.
6. Law Commission, 1989, p 52.
7. Waitangi Tribunal, 1992, pp. 267–68.
8. Waitangi Tribunal, 1992, p. 269.
9. Waitangi Tribunal, *Muriwhenua Fishing Report*. Department of Justice, 1988. p. 219.
10. Waitangi Tribunal, 1988, p. 194; 1992, p. 272.
11. Waitangi Tribunal, 1988, p. 203.
12. Waitangi Tribunal, 1988, p. 204.
13. Waitangi Tribunal, 1992, p. 273.
14. Waitangi Tribunal, 1992, p. 273; Waitangi Tribunal, 1988, p. 194.
15. Law Commission, 1989, pp. 56–7.
16. Waitangi Tribunal, 1988, p. 204.
17. Peter H. Pearse, 'Fisheries Management Regimes for the Future.' Paper presented to XVIIIth Annual Conference of The European Association Of Fisheries Economists, Reykjavik, Iceland, 2007. pp. 1–7. http://www.univ-brest.fr/gdr- amure/eafe/eafe_conf/2007/peter_pearse_eafe2007.pdf
18. Law Commission, 1989, p. 54.
19. Tipene O'Regan, 'Treaty Settlements, Fisheries and the Restoration of Rights.' Thomas Cawthron Memorial Lecture (unpublished). Nelson, August, 1999. pp. 4–5.
20. O'Regan, 1999, p. 5.
21. *New Zealand Maori Council and Latimer v Attorney General* [1987] 1, NZLR 641.
22. Law Commission, 1989, pp. 90–1.
23. Waitangi Tribunal, 1988, p. 211.

MAKING WAVES:
The Waitangi Tribunal

By the early 1960s there was a growing mood for change in New Zealand, particularly from Māori and their supporters. The migration of Māori to the cities from traditional homelands following the Second World War, to work in and enjoy the fruits of an expanding industrial sector, perhaps encouraged that generation to 'get a job and get ahead' and forget the dwindling, sorry state of Māori land holdings and past injustices. However, a new generation of young, urban Māori, educated and aware began a 'Māori renaissance' that started to challenge their elders, the government policies of assimilation and the continuing alienation of Māori land. These groups demanded greater legal and social equality as well as the rejuvenation of Māori culture (including government support for Māori language), and they called on the Crown to end the taking of Māori land. They demanded that the Government elevate the Treaty of Waitangi to its rightful place as the nation's founding document legitimising Pākehā colonisation of New Zealand.

In the lead-up to the 1972 general election, it was the New Zealand Labour Party which most closely represented Māori aspirations. Since becoming the Government in the 1935 elections Labour had dominated the parliamentary seats reserved for the four Māori electorates. After Labour won the 1972 election, led by Norman Kirk, the new government appointed Matiu Rata (MP for Northern Māori) Minister of Māori Affairs. This appointment was important because under National, both the previous Ministers (Ralph Hanan and Duncan McIntyre) had been Pākehā. In fact, Matiu Rata was the first Māori to hold this post since Sir Apirana Ngata had been Minister of Native Affairs from 1928 to 1934.

The status of the Treaty of Waitangi was also elevated to 'the foundation stone of our nation' — a mantra carried on from that time by successive governments. In his book *Maori and the State*, Richard Hill says that Māori urban voices and more conservative Māori Council representatives both insisted that the Kirk Government honour the Treaty, and the Crown 'had no choice but to come to terms with it'.[1] As Minister of Māori Affairs, Matiu Rata was in close touch with various outspoken Māori, such as Dr Patu Hohepa, Dr Ranginui Walker and Titewhai Harawira, as well as with members of the New Zealand Māori Council, such as Graham Latimer. One initiative that Labour believed was key to tempering the insistent and strong Māori voices was to establish a forum where grievances against the Crown and calls for compensation for past losses by Māori could be heard and, it was hoped, resolved. In the end, given the growing awareness among non-Māori (and Māori) of these grievances, thanks to the impact of books, television, public lectures or the activities of groups such as Ngā Tamatoa and the Māori Organisation on Human Rights, Matiu Rata was able to raise sufficient support within the Labour caucus to bring about the establishment of the Waitangi Tribunal. Dr Patu Hohepa (formerly Matiu Rata's advisor) recalls that:

When Matiu became Minister of Māori Affairs he had an illness bout, and asked Norman Kirk to persuade me to take a year's leave to sort out two things:

Norman Kirk and Matiu Rata at Hairini Marae, Bay of Plenty, c.1971. Matiu was appointed Minister of Māori Affairs and Lands in the Kirk Labour government. Both men were instrumental in establishing the Waitangi Tribunal.

NZ Archives: AAWV 23583 21 (R22940968).

Judge Eddie Durie, Waitangi Tribunal,
Wellington, February 1995.

Newspix.co.nz: NZ Herald

[the] 1967 Hanan Maori Affairs Act, and write some ideas for a Māori tiriti [Treaty of Waitangi] Act. I wrote the basic material for Matiu, Norman Kirk insisting I do so urgently. It took some six months to encircle everything concerning Te Tiriti, then report and set up . . . the Tribunal legislation adopted by Labour.[2]

Sir Eddie Durie recalls meeting with Matiu Rata:

I was practising in Tauranga when I first met Matiu because of some clients for whom I was acting. They had contacted Brownie Reweti, the local MP, for help and he called in Matiu. Matiu mentioned to the group the idea of having some body to look at Māori land grievances. At that time he had set up the Māori Land Boards to investigate the same in their districts. We talked of the Māori land taken in Mount Maunganui for a rifle range that the Māori Land Board at Hamilton was investigating, the proposals of the Local Government Commission to include Matapihi Peninsula Māori lands in Tauranga City boundaries, and the proposals to build the Tauranga Harbour bridge and take out part of Whareroa Marae. It was agreed that I would appear for the owners before the Commission and would write a submission on behalf of Whareroa Marae relating to the bridge. There was a successful outcome on both counts. In further talks Matiu mentioned that the task seemed too large for the Māori Land Boards, and that a national body was needed with a permanent, specialist body, but he did not discuss it in any detail. Maybe his thoughts had not taken shape at that time.[3]

Sir Eddie Durie also recalled that later, in August 1974, when he was appointed as a Judge of the Māori Land Court in Rotorua, the then Chief Judge of that court, Kenneth Gillanders-Scott, was assisting with the drafting of a Bill for Matiu Rata for what would become the Waitangi Tribunal legislation:

When Matiu visited Rotorua he organised for me to go to northern Canada, near Alaska, where Judge Thomas Berger of the Canadian High Court was conducting an inquiry into the McKenzie Valley Oil Pipeline and its impact on Inuit and Dene Indians. I had the impression that Matiu thought this new body would operate along similar lines, that is, a body that would meet with the Native Inuit and Indians on the trap lines and at the Spring Camps, rather than in courtrooms. Alas, that was not to be when the first appointments to the Tribunal were made by the National government (Labour went out just after the Act was passed, but before appointments were made). The National government was tardy in making appointments, finally coming up with Ken Scott, Graham Latimer and Laurie Southwich QC, who sat beneath the grand chandeliers of the Intercontinental Hotel ballroom in Auckland. However, when Ben Riwai Couch became Minister of Māori Affairs, he brought me in as Chief Judge and chair to replace the retiring Ken Scott. That was in 1980. He seemed to be aware of Matt's vision and to quietly agree with it, although publicly he held to the line that he was a New Zealander first and a Māori second. Ben had come in on the European ['General'] seat of Wairarapa and needed to keep in tune with his electorate.[4]

Ben Couch would serve as Minister of Māori Affairs from December 1978 until July 1984, when Labour again won office and David Lange became Prime Minister.

By August 1974, when Norman Kirk passed away and the less charismatic Bill Rowling became Prime Minister, the Māori land rights movement had been partially successful with the setting up of the Waitangi Tribunal. However, iwi around the country were impatient with the slow progress being made by government and local authorities in addressing the mixture of historical and contemporary issues, mainly to do with land rights. One particular dispute

OPPOSITE PAGE

Māori Women's Welfare League Conference, Auckland – probably 1955. Third from left: Whina Cooper, who became the League's first president when it was formed in 1951. Far right: Mira Petricevich, later Mira Szaszy, Assistant Dominion Secretary. Both Mira Szaszy and Whina Cooper were closely involved in Māori rights campaigns and the fisheries rights struggle.

ATL: F-40535-1/2. T Ransfield

was to create the spark for a fire of discontent.[5] A 30 kilometre strip of coastal land belonging to Ngāti Wai, located north-east of Whangarei, was designated public reserve by the Whangarei County Council. Ngāti Wai set up a group to lobby for the protection of their land. The group was supported by Whina Cooper, an outspoken leader well-known to many as the first national president of the Māori Women's Welfare League. At the age of eighteen she had taken direct action against a local farmer when he began to drain mudflats which were a well-used source of seafood near her home at Panguru, in the far north of New Zealand. Whina and others obstructed the farmer, who was forced to abandon the operation.

By 1975, many varied land disputes were being actively contested by a range of Māori groups. These were brought together for discussion at a hui held at Te Puea Marae in Mangere, Auckland.[6] Whina Cooper chaired that hui and from the wide range of issues and groups, there emerged angry demands for something different to be done that would attract attention and get results. The suggestion of a protest hīkoi (march) to Parliament, beginning at Te Hāpua in the far north, was adopted. Te Roopu o te Matakite was formed to organise the march and an accompanying petition. Whina Cooper was elected as its leader. Aroha Harris likened the hīkoi to the one carried out in 1867, when 'the remarkable military leader', Riwha Titokowaru, led a hīkoi for peace through the Taranaki and Whanganui districts, following the land wars and land confiscations there.[7] Te Roopu o te Matakite galvanised nation-wide support for the march, an achievement which Aroha Harris described as being 'as great a feat as the march itself'.[8] Matiu Rata made his support clear by sending a telegram 'expressing his good wishes'.[9]

A 'Memorial of Rights' was drawn up, calling for all remaining Māori land to be protected in perpetuity and for all laws that allowed the taking of Māori land to be rescinded. This was signed by over 200 rangatira representing all the tribes of the country as the hīkoi progressed.[10] A separate petition was also carried by hīkoi organisers and some 60,000 people signed this as the hīkoi wound its way to Wellington.[11] The march ended on the steps of Parliament on 13 October 1975. The petition and Memorial of Rights were presented to the Government. The Leader of the Opposition, National's Robert Muldoon (soon to be Prime Minister), was also on the steps of Parliament to greet the protestors. The march was described as 'testament to the depth of feeling about the land issue'.[12] Afterwards, a number of other protest actions were carried out by other Māori groups throughout the country, leaving majority Pākehā New Zealand in no doubt about Māori sincerity and determination to achieve justice on a whole range of issues.

Roughly a month after the land march had begun in Te Hāpua, Matiu Rata and his Cabinet colleagues passed the Treaty of Waitangi Act 1975. Yet as liberal and 'progressive' as they claimed to be, the Labour government would not agree to the Minister's request that the Waitangi Tribunal address Māori grievances going back to 1900. They did agree that the Tribunal would provide a legal process for investigating Māori claims of breaches of the Treaty, and would make

OPPOSITE PAGE

Approximately 3000 Māori land marchers cross the Auckland Harbour Bridge, 23 September 1975. They were heading for Te Ūnga Waka Marae. The march began at Te Hāpua in the Far North on 14 September and ended outside Parliament Buildings in Wellington on 13 October. The hīkoi galvanised Māori opposition to a seemingly endless loss of Māori land, leading to further protests against Crown policies and practices that enabled Māori land to be taken. The hīkoi was pivotal to Māori expectations and politicisation around the Crown investigating redress for long-standing grievances, including fishing rights.

Newspix.co.nz: 1036948 NZ Herald

recommendations to the government from the time the Act came into force. They were hiding behind what Richard Hill describes as 'spurious advice that historical material was lacking'.[13] However, the Waitangi Tribunal could make findings of fact and make recommendations to the Crown, and it was 'the only official body with the authority to determine the meaning and effect of the Treaty of Waitangi, taking into account both its English and Māori versions. The National Party did not oppose the Act although one of its MPs commented that 'the tribunal's responsibilities appear minimal'.[14]

Although the Tribunal was unable to investigate the significant grievances of past years, and politicians felt that they had helped calm the increasingly troubled waters, the Tribunal itself was not about to roll over. Under its new Chairman, Eddie Durie, and despite its limited powers, there were three significant cases (the Manukau Harbour, Motunui and Kaituna River inquiries), in which the Tribunal made findings that were to have a major effect on the Māori fisheries story. In 1985 the Tribunal (consisting of the Chairman, Eddie Durie, and members Sir Graham Latimer and Paul Temm QC) reported on a claim on behalf of the people of the Manukau Harbour. It concerned, among other things, pollution of seafood resources and loss of surrounding land from confiscations after the New Zealand wars and for public works.[15]

The matter of Māori fisheries had earlier been important in both the Motunui and Kaituna River claims. In the former, Te Āti Awa of Taranaki (spokesperson, Aila Taylor, and other witnesses) gave evidence that their tradition of gathering shellfish on offshore reefs was essential to providing hospitality and for tribal mana. The Taranaki synthetic fuel factory, which was partly Crown-owned, planned to discharge untreated effluent into the sea and thereby contaminate shellfish beds that had been under tribal control for centuries. The Tribunal condemned the planners for failing to consider the Māori cultural approach to water as a source of food. In the second case, Māori claimants objected to a proposal to discharge Rotorua city's treated sewage into the Kaituna River, an important eel fishery and, at its estuary, a significant shellfish gathering area. The scheme was stopped as a result of the Tribunal's findings.[16]

In the Manukau case, the Tribunal took a more detailed look at the fisheries of the harbour and how these were important to Māori. It noted in its report that:

> From 1894 to the present day the fisheries laws have provided that the provisions of the Fisheries Act shall not affect 'any existing Maori fishing rights' but the Courts have held that save from some special provision in an Act or Crown Grant, there are no 'existing Maori fishing rights'. This has thus become an empty provision. Those words mean nothing . . .
>
> In 1894 there was provision for oyster beds to be reserved for Maoris. Only one Maori Oyster Reserve survives in the Manukau but the oysters no longer survive there. From 1903 to 1962 there was provision for the reservation of tribal fishing grounds, but this was dependent on the exercise of a Ministerial discretion and no reserves were created. The current Fisheries Act [Fisheries Act 1983] enables the Director-General of Fisheries to confer 'special rights' on 'special communities' in respect of defined sea areas. Thus 'rights' now depend upon the exercise of an administrative discretion and there is now an open reluctance to refer to Maori fishing grounds or the Treaty of Waitangi, or to confer any priority on the Maori community in terms of that Treaty.
>
> We think it would be unfortunate if Maori fishing rights fell to be determined solely on a literal interpretation of the Treaty which guarantees

OPPOSITE PAGE

Joseph Cooper, son of Māori land march organiser Whina Cooper (aged 83), presents the land rights scroll proclaiming the Māori case for a return of their lands to the Prime Minister, Bill Rowling, at Parliament, Wellington, 13 October 1975.

Fairfax NZ: Waikato Times *archive 120712*

an exclusive use of all Maori fisheries, for Maori fisheries are extensive and indeed, the whole of the Manukau could be described as a traditional Maori fishery . . . There is obvious potential for conflict between Maori, private and commercial fishing interests and the potential for conflict should be minimised. Compromises will be necessary. But the answer is not in the blatant denial of Maori rights, it is not in glossing over the problem, and it is not in the maintenance of a Fisheries Act that contains empty words and clearly fails to match the promises of the Treaty. Those answers merely strengthen, and probably cause, Maori demands for the ownership of harbours, and exclusive fishing grounds, demands based upon a strict interpretation of the Treaty. Instead a genuine search should begin to define the options available for the recognition and protection of Maori fishing grounds and securing compensation for Maori fishing losses.

In the meantime pending the formulation of better policies and better laws to honour the fishing guarantees of the Treaty we consider that the Whatapaka and Pukaki inlets should be reserved as requested. The Treaty is simply so clear and the lack of research for better alternatives is simply so obvious that we have no alternative but to accede to that demand.[17]

The Tribunal also noted that the Ministry of Fisheries had already developed proposals to deal with the problem of overfishing throughout the country by introducing individual transferable catch quota systems and fishery management plans.[18]

This is the first time the Tribunal had made the point strongly that the Treaty guaranteed Māori 'rangatiratanga' over their fisheries and that the Crown had ignored these rights. The Tribunal made no real attempt to define those rights, but observed that the entire Manukau Harbour could be described as a 'traditional Māori fishery'. This was a very tentative foray into the world of Māori fisheries, a nibbling of the bait, for the prize was that Article Two of the Treaty guaranteed Māori their fisheries for as long as they wished to retain them.

The 1984 Lange-led Labour government had given the Tribunal the power to investigate claims from 1840, the date of the signing of the Treaty of Waitangi. The number of Tribunal members had been increased from three to seven, at least four of whom were required to be of Māori ancestry. Another seven were to be appointed as deputy or alternate members. The Tribunal had also gained research and administrative staff. However, it remained an advisory body, with no power to enforce its recommendations.[19]

Under the tutelage of Eddie Durie, the Tribunal showed itself capable of being more than a commission of inquiry. Hearings developed, even with the limited resources at its disposal, into user-friendly inquiries — something Māori claimants had seldom before experienced. Claimants were given enough opportunity to demonstrate their case, with site visits, convening at claimant marae, and exploring in depth claimant history and background, that the actual grievance was seen in a more holistic way, throwing light on the interconnections between Māori claimants and their physical and social environment. The Tribunal also took care to take account of lore

OPPOSITE PAGE TOP

Minister of Māori Affairs Matiu Rata, at the pinnacle of his political career, meets Māori land protesters on the steps of Parliament, Wellington, 1975. Matiu's empathy was with the protestors, but as a part of Bill Rowling's Labour government, he had to appear neutral.

Fairfax NZ: parl_256453 1975 Dominion Post 10/75

OPPOSITE PAGE BOTTOM

The audience at Manukorihi marae (Waitara) attending the announcement of the Waitangi Tribunal decision on protecting North Taranaki fishing resources, 2 April 1983. Over 300 people from around the North Island gathered there to discuss the government's rejection of Waitangi Tribunal recommendations on fishery protection.

Fairfax NZ: TDN-Maoriculture3-0075A Taranaki Daily News

(customs and traditions) as well as law in its 'findings'. For example, the Tribunal hearings into the Mangonui Sewerage Scheme and Muriwhenua claims visited several sites of importance to the claimants. Shane Jones recalls the Tribunal visit to Te Rerenga Wairua (Cape Reinga), the legendary leaping-place of the spirits where, Māori believe, the spirits of the dead depart the island to return to the spiritual homeland – Hawaiki. While members of the Tribunal stood on the rocky pathway on the steep ridge running down to the foaming water's edge, a huge shoal of araara (trevally) darkened the boiling seas, suggesting to several witnesses an omen for the future.[20]

The next major leap forward for Māori fisheries began with the Tribunal's Muriwhenua hearings. The Tribunal was indeed, intended or otherwise, going to 'make waves'.

OPPOSITE PAGE

Members of the Waitangi Tribunal, 25 May 1985. From left: Sir Graham Latimer, Chief Judge Eddie Durie and Paul Temm QC. Under Eddie Durie's chairmanship, the Tribunal took care to thoroughly familiarise itself with the claimants and their circumstances during the hearing process.

Newspix.co.nz: 1080161 NZ Herald

ENDNOTES

1. Richard Hill, *Maori and the State: Crown–Maori Relations in New Zealand/Aotearoa, 1950–2000.* Victoria University Press, 2009. p. 164.

2. Patu Hohepa, pers. comm., 2014.

3. Edward Taihakurei (Eddie) Durie, pers. comm., 2014.

4. Edward Taihakurei (Eddie) Durie, pers. comm., 2014.

5. Aroha Harris. *Hīkoi – Forty Years of Māori Protest.* Huia Publishers, 2004. p. 68.

6. Harris, 2004, p. 70.

7. Ibid.

8. Ibid.

9. Ibid, p. 75.

10. Ibid, p. 74.

11. Ibid.

12. Ibid, p. 76.
13. Hill, 2009, p. 165.
14. http://www.teara.govt.nz/en/waitangi-tribunal p. 1.
15. http://www.teara.govt.nz/en/waitangi-tribunal p. 2.
16. http://www.teara.govt.nz/en/waitangi-tribunal p. 2.
17. Waitangi Tribunal, 1983, pp. 80–4.
18. Waitangi Tribunal, 1983, p. 83.
19. http://www.teara.govt.nz/en/waitangi-tribunal
20. Shane Jones, pers. comm., 2015.

AN OCEAN WINDSWELL:
Māori objections to government actions over fisheries

In the late 1860s, an inexorable trend towards State control over Māori fisheries began, starting when over-fishing in particular areas became apparent. From that time, as fishing techniques improved and fish consumption and exports grew, the Crown began to set rules to control fishing. Any rights that Māori had were ignored. Māori participation in fisheries became more and more individualistic. Tribal authorities became increasingly dismayed at being ignored and even discriminated against by new government fisheries legislation. As noted earlier, the first legislative intervention in 1866 in the form of the Oyster Fisheries Act was passed without any consultation with Māori. The Crown assumed that it had the right to restrict or deny access to fisheries by both Māori and non-Māori alike. As the Waitangi Tribunal has noted, 'This intervention generated a continuous stream of complaints and protest by Maori in various parts of the country during the remainder of the nineteenth century and down to the present time'.[1]

The Muriwhenua Fishing Report documents many of the protests by Māori in the last three decades of the nineteenth century and into the twentieth century. The Waitangi Tribunal did not confine its discussion solely to grievances and protests by the Muriwhenua tribes, but extended it to those by Māori throughout New Zealand.

These complaints and protests took a variety of forms. Some were made direct to the Native Department or the Native Minister. Others took the form of petitions to the House of Representatives. The Muriwhenua Fishing Report records in appendix 8 that some 46 Maori fishing petitions were referred to the Native Affairs Committee in the period 1869 to 1910. Nine of these were from Ngai Tahu. Yet other protests were made by Maori MPs, including H. K. Taiaroa and Tame Parata from Ngai Tahu, during debates on various fishing Bills before the House of Representatives or Legislative Council. Court proceedings by Maori were yet another form of protest. These later became more frequent during the twentieth century.[2]

Unlike many other tribes, Muriwhenua were heavily reliant on fishing for their livelihoods. The government began actively to encourage fisheries around the 1880s, particularly to boost export earnings. Muriwhenua Māori, despite being so dependent on fishing, were not given any special consideration and were unable to participate or exercise their fishing rights generally. Fishing methods changed too, so that larger trawlers were able to work in virtually any part of the coastal seas. By the 1960s over-fishing had become apparent.

How had this situation been able to develop and why had it been allowed by successive governments, despite opposition from Māori? Part of the explanation can be found by referring back to the discussion on the Treaty of Waitangi, and the interpretation put on Māori rights seemingly enshrined there, by the settlers and successive governments.

Crown and settler attitudes to fisheries and fishing

The reasons for the new settlers and successive governments placing little or no value on the Treaty of Waitangi are not clear. The Waitangi Tribunal said:

> The settlers brought as part of their mental baggage the belief that the foreshore and the sea were common to all for the purpose of getting fish. This belief stemmed from the common law doctrine that the Crown is *prime facie* owner of the foreshore. The settlers and indeed successive settler governments simply assumed that the Crown prerogative overrode or qualified the fishing rights guaranteed to Maori by the Treaty.[3]
>
> . . . The Crown appears never to have entertained any doubt about its right thus to assume control over Maori fisheries. Nor is there any evidence that in doing so it believed it should be mindful of the fishing rights guaranteed to Maori by the Treaty. As a consequence no effort was made first to consult with Maori before exercising legislative control over their fisheries. As we have observed this total disregard of Maori fishing rights is in marked contrast to the Crown's attitude to Maori land rights notwithstanding both were protected by article 2 of the Treaty. At least in the case of land the Crown recognised that Maori owned the land and it was necessary to negotiate with Maori for its acquisition. Despite the Treaty guarantee of Maori fisheries the Crown for the most part acted as if it, not Maori, owned this extremely valuable resource.[4]

After analysing the various court cases and legislation, the Waitangi Tribunal commented:

> With the limited exception of the case of Te Weehi [in 1986] the New Zealand courts for the whole of this century have declined, in the absence of express legislation, to recognise either Maori customary fishing rights or Maori rights under article 2 of the Treaty of Waitangi. In so doing they have upheld the Crown arguments which have consistently, for more than a century, denied that any such Maori rights are entitled to legal recognition or enforcement.[5]

The Te Weehi case is discussed in detail later (Chapter 5) because it is pivotal in the Māori fisheries story.

Despite Māori not being consulted by successive governments about fisheries law, why did Māori not get involved in commercial fishing anyway? Were they able to participate equally and perhaps even carry an advantage into commercial activities, with their superior knowledge of fish species, fishing grounds and the like? The Tribunal addressed this question in reference to Ngāi Tahu. It is obvious that the severe impediments faced by Ngāi Tahu also affected other tribes' fishing potential:

> The reason they [Ngai Tahu] were not so involved but confined their activity to fishing for their personal needs and those of their guests is directly attributable to the Crown . . . By serious and sustained breaches of its Treaty obligations the Crown, due to their massive land acquisitions from Ngai Tahu rendered them virtually landless and without any economic base. Given their impoverishment, Ngai Tahu were unable to maintain their commercial fishing activities so evident before the land purchases were completed in the 1860s. It is therefore understandable that having been reduced to fishing for their own personal consumption, their recorded protests and complaints concentrated on inland fisheries with particular concern for access to eels and whitebait. Many invoked the guarantee of their fishing rights under the Treaty. Inland waters were more readily accessible and, scattered as they were across the entire Ngai Tahu territory, they served as a major source of sustenance especially in the inland areas . . .[6]

OPPOSITE PAGE

Eel weir at Pungarehu, on the banks of the Whanganui River, c. 1880. These weirs were effective in trapping the eels that provided basic food for many tribes.

ATL: 1/1-000483-G. William James Harding

Digging for toheroa, Northland, c. 1920.

ATL: 1/1-010576-G. Northwood brothers

In the Muriwhenua Fishing Report, the Tribunal stated that:

> In Muriwhenua 'the tribes gained few of the benefits of trade, including trade in fish, that were available to those tribes closer to the main Pakeha settlements. It was after fish laws were made restricting Maori that complaints began.
>
> Before 1870 non-Maori commercial fishing was mainly rudimentary and posed no competition. Regulation was still unnecessary. Most non-Maori fishing was for personal consumption, save for the commercialisation of the northern oysters by non-Maori in the 1860s. Fish remained plentiful.[7]

As had occurred in Ngāi Tahu and much of the rest of the country, Muriwhenua Māori economic welfare declined markedly after 1870:

> The ravages of war, land loss, population decline and hostility to Maori trading were by far the main causes. These losses caused a large decline in fishing. The acquisition of Maori land by the Crown without due consideration for what should be left to Maori to sustain them; the decline in tribal authority due in large part to the introduction of new land titles, contributed to a reduction of tribal fishing . . .
>
> Parliamentary laws seriously impacted on all Maori fishing. A series of attenuating laws followed almost immediately upon the attainment of a Pakeha majority, the outbreak of wars and the transfer of British control of Maori affairs. The first of these laws (in 1862) related to land. It ended the communal basis of land ownership and destroyed traditional tribal controls. The fish laws related first to oysters (from 1866) when the Maori sale of oysters was disallowed, to freshwater fisheries (from 1867), and finally to all fisheries (from 1877).
>
> The presumptions fixed in those first laws permeate all subsequent legislation. They were that Maori interests should be accommodated by reserving particular fishing grounds for Maori; that Maori fishing has no commercial component and grounds reserved must be for personal needs; that Maori participation in the commercial fishing industry should be on no other terms than those provided for all citizens; no allowances should be made for Maori fishing methods, gear or rules for resource management; the recognition of fishing should be an act of state; only Parliament or a department of state should authorise the reservation of fishing grounds; there should be no provision for the courts to recognise rights on proof of customary entitlement; some acknowledgement should be made of Maori fishing interests by incorporating words of a general nature in fishing laws.[8]

The Tribunal also discussed the Fish Protection Act 1877 (section 8) in terms of its being the first comprehensive fisheries control measure in New Zealand, and the way in which it appeared to recognise the Treaty of Waitangi:

> It recognised the Treaty of Waitangi but the manner in which it did so illustrates a recurring theme, apparent also in Maori land laws (the Native Land Act 1862 for example) that Maori concerns for the recognition of Treaty interests could be met by mentioning the Treaty in the Act, in a general way, and although nearly everything else in the Act might be contrary to Treaty principles.[9]

Discussing possible reasons for this Treaty reference in the 1877 Act, The Tribunal surmised that:

> The more likely event is that section 8 was really window dressing, inserted in the face of claims by Maori members [of Parliament] that the Treaty should be recognised and fishing rights upheld, while no one really knew or very much cared what section 8 entailed.[10] There were a string of subsequent fisheries Acts, amendments, regulations and so on but despite Maori member opposition about

OPPOSITE PAGE

Members of the Ngawati family alongside a pou rāhui used to mark eel fishing territories, Otiria, Northland, c. 1900.

ATL: David Henry Graham, 1/4-018226-G

the only concession to them was the continued inclusion (apart from a short period 1884 to 1903 – after which it was reinstated), of a provision similar to section 8 in the Acts including the Fisheries Act 1983.[11]

It is clear from Tribunal findings in both the Ngāi Tahu and Muriwhenua fisheries reports that from the first fisheries legislation, Māori parliamentarians voiced strong opposition to various parts of these Acts, particularly concerning the non-recognition of Māori Treaty rights. There was a consistent argument advanced by government that Māori Treaty rights were no more than non-commercial rights to harvest.

Outside Parliament, Māori had no option but to challenge the increasingly stifling fisheries laws and regulations. The Tribunal pointed out that Māori complaints against these laws began in about 1879 and continued to the present day (the Muriwhenua inquiry began in 1986). However, this was hugely costly and stultifying for those tribes and individuals concerned. The Crown would not include a specific reference to the Treaty in legislation; this would give powers to Māori over fisheries, and perhaps set precedents in other areas – particularly land and waterways – which the Crown was carefully avoiding.

Not surprisingly, legal challenges to the fishery laws did occur. These are documented in both the Waitangi Tribunal Ngāi Tahu and Muriwhenua fishing reports and in the Law Commission report published in 1989, at the behest of the Minister of Justice at the time, Geoffrey Palmer. Historian Tony Walzl, presenting evidence to the Ngāi Tahu Tribunal hearings, said that Māori protest over lost fisheries and fishing rights began building to a crescendo, and that this forced the Crown to re-evaluate its stance around 1885.[12] Earlier Māori fishery petitions (of the 1870s) had mainly referred to inland waters, the navigation of rivers and the protection of eel-weirs.

One of the first petitions relative to sea-fisheries was submitted in 1879 by Arama Karaka, who had been a delegate to the Orakei Māori parliament. The petition referred to European legislation over fisheries, and demanded 'That the laws of the Europeans should not affect the deep-sea fisheries and pipi-beds, because the Europeans have only a right to the dry land.'[13] At that same Orakei Māori parliament, several participants complained that the Crown was taking over Māori fisheries, contrary to guarantees in the Treaty.[14] Earlier, in 1876, Wi Tako (Wiremu Tako Ngatata, who in 1872 became the first Māori member of the Legislative Council) objected strongly to the government introducing legislation designed to control fishing. He stated that Māori fisheries were secured by the Treaty of Waitangi and was adamant that the government had no right to interfere in them.[15]

Petitions to Parliament continued to be made by Māori, but the precedent set in the 1914 Waipapakura court case effectively stifled any hope of success. This was one of the two court cases which stand out as pivotal to understanding the role that the courts played in the Māori fisheries story. The other was the Te Weehi case (discussed below and in the next chapter).

Waipapakura, a Māori woman from Waitara, had been catching whitebait in August 1913. Fisheries Officer Hempton had confiscated her nets because she was deemed to have been fishing illegally. Waipapakura challenged the fisheries officer in court, claiming that she was exercising her right as a 'Native' and was not therefore subject to the Fisheries Act. After losing in the lower court, Waipapakura resorted to the Court of Appeal. The Solicitor General, J. W. Salmond, appeared for the Crown, indicating that the government was very concerned with winning the case.[16] The Chief Justice, Sir Robert Stout, ruled that the court was bound by an earlier opinion of the Privy Council that the Treaty of Waitangi conferred no legal rights without legislative enactment, and that section 77(2) of the Fisheries Act 1908 created no rights either, but merely protected those Māori fishing practices specifically provided for by statute. The Crown therefore won its case and Waipapakura lost her whitebait nets, although the Crown was ordered to pay court costs.

Thus it was that for the period 1900–1987, no general right of Māori fishery was recognised at law, until the Te Weehi case in 1986. In the Tribunal's view, 'Even that case may have limited scope, for while the fishing industry has expanded enormously, the customary right referred to related to gathering for personal needs in a customary way.'[17] The Te Weehi case is discussed in detail in the next chapter. It is thought of today as pivotal because of its timing, the compelling case brought by Tom Te Weehi, and the ensuing judgement, which could be described as visionary.

From the 1950s, growing concern was expressed by the New Zealand Māori Council and Māori leaders about the depletion of the fisheries.[18] The Waitangi Tribunal noted that at a hui held in 1985 to discuss fisheries:

> Maori thinking had changed little from that of the rangatira who assembled at Orakei in 1879 [a meeting of the Maori Parliament]. Instead of identifying particular fishing grounds for reserves, tribes were defining their boundaries and claiming fishing rights along entire coasts: 'It was control that was mainly talked of, not the exclusion of the public. Once again, it was a matter of mana.'[19]

The Tribunal recorded opposition to fishing laws in 93 petitions, 'most on behalf of many persons, one on behalf of 11,976. In most of them, the Treaty of Waitangi was invoked, and more often than not, specific fish laws were alleged to be contrary to the Treaty.'[20]

> Much effort was spent in seeking the reservation of particular fishing grounds. Laws enabling such reserves applied from 1900 to 1962. We can find no reserves ever established under those provisions. A very few reserves were gained after petitioning Parliament. Apart from the provisions for fishing reserves, and the occasional statutory conferral of Maori fishing rights in particular areas, Maori fishing interests were recognised in a provision for special licences to gather seafood for hui, a provision in the 1930s for Maori to receive a lesser unemployment benefit due to their access to natural resources. The special provisions for fishing reserves or the capture of seafood for hui converted substantive rights to mere privileges. It encouraged the view that allowances for Maori generally ought not to be made.
>
> . . . In the period 1900 to 1980s Maori commercial fishing in Muriwhenua [as well as in other areas] was characterised by individual effort, mainly in areas less affected by land loss and with substantial Maori populations, usually on a part-time basis, and often by low income farmers and agricultural workers seeking to supplement their wages or returns . . . State enforcement action against Maori exercising their claimed Treaty right to fish continued to apply and to dissuade Maori from their usual fishing activities. State laws compelled Maori dependent on fishing to operate as individuals within the state fishing regime. Restricted licensing from 1937 to 1963, for the conservation of the resource, impacted on Maori by reducing the available licences and restricting their methods and gear.[21]

A significant expansion in the industry followed delicensing in 1963, when concessionary loans and export incentives were made available from government. Overfishing in the inshore fishery was apparent in the late 1970s and some stocks became seriously threatened. Further government assistance was given to exploit the offshore fishery. Severe overfishing led to the forced retirement of a number of small-time commercial fishermen in the early 1980s. Eventually it gave rise to the Quota Management System.[22]

In both the Ngāi Tahu and Muriwhenua claims to the Waitangi Tribunal the tribes were adamant in their assertion of their fisheries rights. Ngāi Tahu stated that:

> . . . article 2 of the Treaty guarantees protection of tribal tino rangatiratanga over their fishery and that there can be no restriction of any kind on the exercise of that rangatiratanga in the seas off the

Ngai Tahu coast line. Accordingly, Ngai Tahu claim ownership of the entire marine fishery adjacent to their tribal lands, including all property and user rights, commercial and otherwise, inherent in the business and activity of fishing. They claim the absolute right to fish in those waters without any restriction whatsoever by whomsoever and claim further that they are entitled to have that right protected by the Crown.[23]

Ngāi Tahu and Muriwhenua strongly protested that they had been dispossessed of a resource: 'Pollution and over-fishing were mainly responsible for the loss. Sewage outlets, industrial discharges, river water deviations and agricultural run offs have contaminated and depleted traditional beds and fishing grounds.'[24] The Tribunal concluded that:

> . . . while, as is to be expected, the Ngai Tahu sea fisheries differed from those of the Muriwhenua tribes, both were actively and regularly engaged in the business and activity of fishing. Both exercised rangatiratanga over the fisheries in the waters off their respective rohe . . . their rangatiratanga encompassed the whole of the sea territory extending 12 miles or so from their lengthy coastline notwithstanding that, for various reasons, not all of it was fished or that some parts were fished more intensively or extensively than others.'[25]

Māori resilience and opposition continue

This chapter continues the story of Māori resilience in the face of mounting odds stacked against them. Once the Treaty had been signed, successive governments essentially consigned it to history and assumed powers over land and fisheries that were in direct conflict with the Treaty, particularly Article Two. As the settler population grew, Māori were outnumbered by about 1870. This swelling population and the growing cash economy provided opportunities for Māori to expand their trading in fish to help feed the growing demand. However, pressure on local fisheries, starting with the desirable oysters, spurred the government to pass laws restricting the harvest. Similarly, supposedly for reasons of conserving the fisheries, other fish species were brought under legislative control. A succession of fisheries Acts and regulations were passed by Parliament without any consultation with or agreement from Māori; to the contrary, Māori members of Parliament voiced strong opposition to many aspects of this legislation.

Outside Parliament, from the 1870s on, Māori tribes and individuals also objected to aspects of the new fisheries laws, organising petitions and taking direct actions. However, these had little effect, as the Crown continually pointed to Court rulings stating that unless the Treaty provisions were included in legislation, they could have no effect. Until the 1980s, the Crown made no effort to remedy that injustice. But the tide of Māori resistance continued to rise. In the next chapter, the story begins with the government changing the way commercial fishing is managed. Other important events, analogous to a major earthquake offshore, allowed for Māori to surf the subsequent tidal wave that brought the Māori fisheries Treaty entitlement to the shore.

OPPOSITE PAGE

Mr H Nutira, checking eels drying on the banks of Te Roto o Wairewa (Lake Forsyth), Central Canterbury, May 1948. Mrs Nutira is seated in the centre of the group of women weaving.

ATL: F-40047-1/2. KV Bigwood

ENDNOTES

1. Waitangi Tribunal, *Ngai Tahu Sea Fisheries Report*. Department of Justice, 1992. pp. 160–61.
2. Ibid.
3. Waitangi Tribunal, 1992, p. 174.
4. Ibid, p. 175.
5. Ibid, p. 180.
6. Ibid, p. 213.
7. Waitangi Tribunal, *Muriwhenua Fishing Report*. Department of Justice, 1988. pp. 220–21.
8. Waitangi Tribunal, 1988, pp. 221–22.
9. Ibid, p. 85.
10. Ibid.
11. NZ Law Commission, *The Treaty of Waitangi and Maori Fisheries*, Prelim. Paper No. 9, 1989. pp. 55–6. See also Waitangi Tribunal, 1988, pp. 86–9, for detailed discussion.
12. Waitangi Tribunal, 1992, p. 162.
13. Ibid.
14. *Press*, Volume XXXI, Issue 4239, 28 February 1879, p. 5.
15. *Press*, Volume XXVI, Issue 3403, 31 July 1876, p. 2.
16. *Evening Post,* 31 July 1914, p. 11.
17. Waitangi Tribunal, 1988, p. 99.
18. Waitangi Tribunal, 1992, pp. 211–12.
19. Ibid, p. 212.
20. Ibid, pp. 330–33.
21. *Muriwhenua Fishing Report*. p 222.
22. Waitangi Tribunal, 1988, pp. 222–23.
23. Waitangi Tribunal, 1992, p. 1.
24. Ibid, p. 9.
25. Ibid, p. 113.

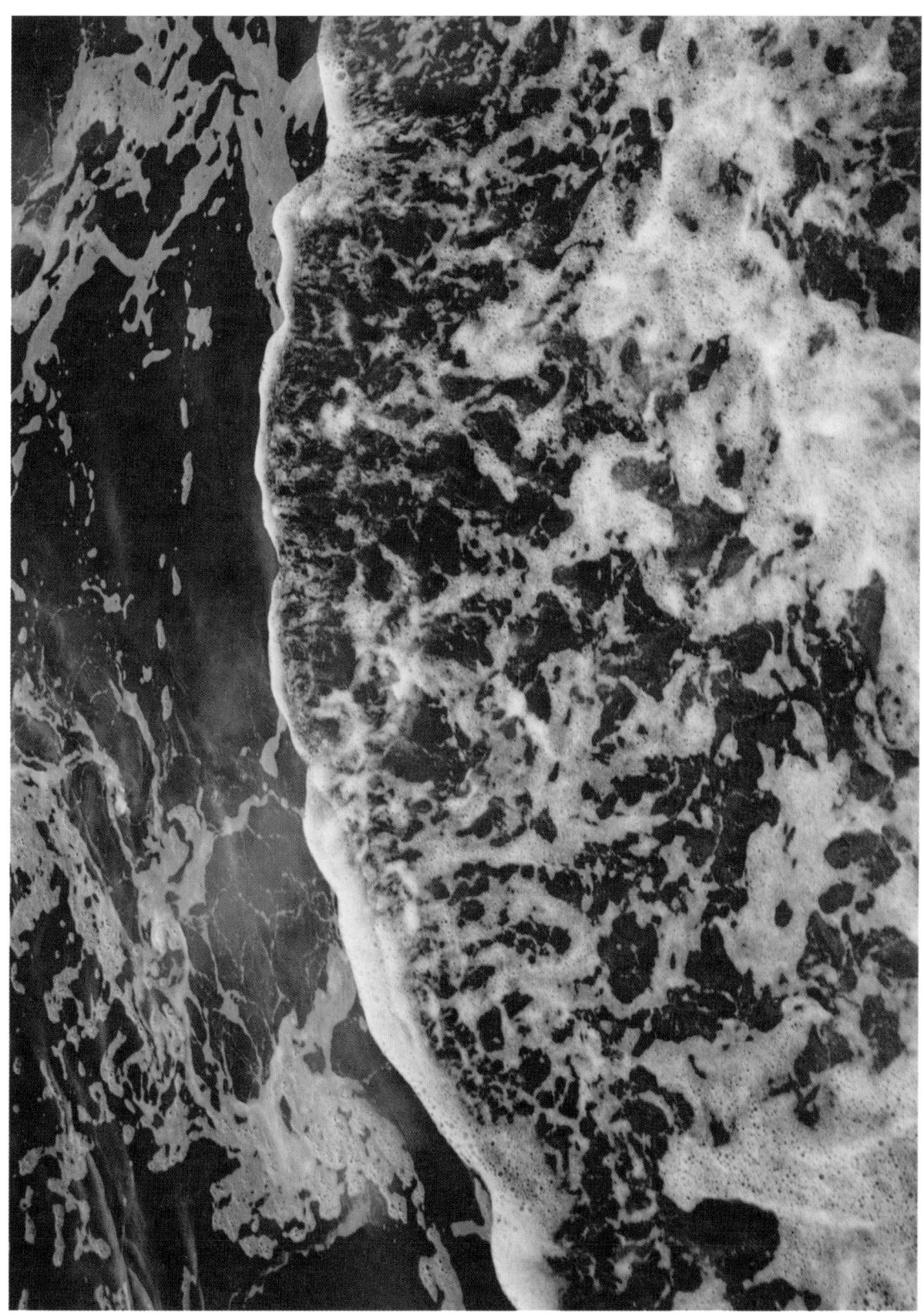

AN INCOMING TIDE:
Government fisheries legislation in the 1980s

By 1980, a succession of fisheries laws had brought continued calls from Māori MPs, tribal representatives and individuals for the Crown to recognise Māori rights to their fisheries. However, all the costly court action and other legal and physical challenges had little or no impact. Māori lost their court cases, their petitions to Parliament were declined, and their politicians were ignored. The Crown continued its control of fisheries through legislation and regulation despite Māori protests and challenges. It proved highly successful both in defending proceedings brought by Māori to assert their Treaty fishing rights and in prosecuting Māori for alleged fishery offences. Throughout, the Crown held firm to its position that the common law right of everyone to fish in streams, estuaries, lakes and the sea over-rode Māori Treaty rights. The Crown was well aware that they could have altered this position and given proper effect to what was agreed in the Treaty of Waitangi, or at least consulted Māori before passing laws to control fishing and protect fisheries. Not only did the Crown over-ride Māori opposition, but there was an almost anti-Māori or racist element to the legislation that, in several instances, allowed individuals and companies rights to fisheries denied to tribes and hapū. These actions made it clear to Māori that the Crown viewed its rights to be paramount. Nevertheless, Māori voices were incessantly raised.

In 1982, after years of mounting pressure and declining yields for both inshore and deep-sea fisheries, and appeals to the government from recreational and commercial fishers alike, the government acted under the leadership of Prime Minister Robert Muldoon. A moratorium was imposed on new entrants to the inshore fishing industry while experts investigated the nature and extent of the problems. The outcome was the passing of the Fisheries Act 1983. This 'placed heavy emphasis on the need to conserve and enhance depleted resources and to bring a greater measure of economic security to those in the industry . . . The principal measures designed to achieve these objectives were to impose fishery management plans for fishery areas into which the seas around New Zealand were divided. The new law was purposely restrictive and provided for controlled fishery areas to be declared in order to fix quotas on fish species, on persons allowed to fish the area and fishing had to be licensed'.[1] These were the seeds of the quota management system.

In an interview, Doug Kidd[2] recounted a fortuitous brush with the (fisheries) law in 1983:

> I substituted on the Agricultural sub-committee [of Parliament] for a meeting and they were in the consideration phase on the consolidation of the 1908 Fishing Act. Such hearings that had been held – and there weren't many – and so there is the parliamentary council, officials from the agriculture and fisheries . . . Lands and Survey. Going back to all the fisheries Acts 'nothing in this act should affect Māori fishing rights', no definition, nothing, their phrase. 'Oh we'll get rid of that', they said, 'it's redundant, it's obsolete, what the hell does it mean anyway'. And I said to the chairman, 'No, we can't do that', and I gave them painful lessons on why.

I told the story of the marae net at Ohau Beach (when communal drag netting occurred with the fish distributed to the families in the area) and what happened with the roads down in Marlborough (where Maori land rather than private land was routinely taken for roading) and said, 'you have no idea what would happen if we take that out'. The officials sitting there dumb, didn't have anything to say. Happy to go along, they weren't being wicked, or racist or anything, they were just people of their times. As we all get very sensitive and sorted and head off into the new sunny up-lands we need to remember they weren't bad people, they just had no idea and were just brought up that way. So anyway I can honestly say that I saved the Māori fishing rights. The Bill comes back to the house and the fisheries legislation never got any debate because no one understood anything about it, most MPs never did, still don't. When I was Minister of Fisheries for two terms I hardly ever got any representations from MPs.

The Quota Management System

The Quota Management System had its origin in work by economists and resource managers. A proponent of the system, Professor Peter Pearse, outlined the rationale for it in 2007. The professor described the earlier world fishing scene — when fish were plentiful — as 'open access':

> From an economic viewpoint, the open access regime was natural and appropriate as long as the fish stocks could easily satisfy all the demands on them. The open access policy was supported by the generally accepted view that ocean fisheries were inexhaustible. But as the world population expanded and fishing fleets grew and employed new technologies for catching fish there was a need to control fishing throughout the world . . . So, in the twentieth century, governments of fishing nations launched a massive regulatory effort to protect stocks from overfishing. These efforts did not change the fundamental rights of fishermen, but government regulators progressively introduced closed fishing seasons, closed areas, and myriad restrictions on boats and fishing gear in an attempt to constrain catches to the sustainable productive capacity of the stocks. But given continued expansion of fishing fleets and advances in technology, these efforts had mixed success. After World War II it was clear that many of the world's most valuable stocks were fully exploited, and many were being depleted. Something had to be done to satisfy the need to conserve fish species for future generations but allow current fishers to remain economic. Economists became involved and devised the principles of the quota management scheme.

> In the 1960s governments introduced a permitting system that limited the number of fishing boats. Thus access to commercial fishing was only by way of obtaining a permit. However, fishers got around these measures by building larger boats and using more sophisticated technology. So, in the late 1970s, the idea of allotting each licensed fisher a specific share of the total allowable catch was introduced. The idea was that if license holders had the right to a defined share they would no longer be driven by the wasteful competition for the catch, which was behind both the conservation problem and the economic problem. With the security of defined shares of the catch, and the ability to adjust them through purchase and sale, fishers restructured their operations to achieve economies of scale and harvest their quotas as efficiently as possible, or alternatively sold them to others. The excess capacity of fleets was soon eliminated by the fishers themselves. In addition, now freed from the hectic race for the fish in brief fishing seasons, fishers could fish for most of the year, when markets were most favourable, and take time to clean and prepare their fish for the best prices. Fishers also found they had a common interest in cooperating to protect the stocks, sustain and increase the total catch, enforce the rules of fishing and generally to improve management — not just to protect their catch but also to increase the value of their assets in fishing rights.[3]

The situation described by Pearse was similar in many respects to that which played out in the New Zealand fisheries economy:

> By 1980, the overfishing of the inshore fish stocks was abundantly clear, and the time for re-appraisal had arrived. Drastic action seemed necessary to reduce the fishing effort and the small and part-time fishermen were the first to be removed[4]. . .

This was brought about by the Fisheries Act 1983. Importantly, the new Act had retained a reference to Māori fishing rights – as had previous legislation. The specific reference was s88(2): 'provided that nothing in the Act shall affect any Maori fishing rights.' Nevertheless, it did affect them drastically.

A key aspect of the new law was that 'commercial fishermen' were defined. These were deemed to be people fishing commercially most of the time. Part-time fishers were excluded. The restricted definition of 'commercial fishermen' caused much distress among small-time fishers, many of whom were Māori, and who overnight were excluded from any further commercial fishing. Fishing boats were to be registered and a permit was required. Permits could be restricted to specific areas, fish species, quantities, methods, types of fishing gear and periods of time, as fixed by the Director-General of the Ministry of Agriculture and Fisheries.[5]

> Maori objected but to no avail. For the Muriwhenua people it meant simply that the whole of 'their' fisheries were invaded, as both small and large operators worked their way north to the extensive Muriwhenua coastline . . . It was essential for the Muriwhenua people that they should seek to survive as individual fishermen within the alternative fishing regime that was imposed. Their livelihoods, families and communities depended upon their doing so. In the time honoured way however, they were part-timers, sharing commitment between their ancestral lands and seas.[6] So it was that the Muriwhenua Maori lost not only 'their' fish to outside fishermen, as the grounds they had nurtured for centuries were largely fished out, but their fishing livelihoods too, and their ancient association with the seas was virtually ended.[7]

The situation facing Muriwhenua tribes was replicated throughout New Zealand to a greater or lesser extent. The Waitangi Tribunal Muriwhenua hearings, which began in December 1986, demonstrated that allowing commercial fishing by outsiders interfered with the Treaty guarantee that Muriwhenua and Māori people generally have 'full, exclusive and undisturbed' rights to maintain and develop their fisheries as they saw fit. This was a clear breach of the Treaty.[8] However, worse was to follow.

Introducing the Quota Management System in New Zealand

The forerunner to the Quota Management System (QMS) had been introduced in 1983 primarily to control seven deepwater fish stocks. At the time, the deepwater fisheries stocks were relatively healthy; this system was implemented to prevent over-fishing or over-capitalisation from occurring as it had in the inshore fishery. Proponents of a full Quota Management System (QMS) and individual transferable quotas claimed that there were large efficiency gains to be made by using the methodology.

In 1983 came the publication of a document entitled *Future policy for the inshore fishery – a discussion paper*. This summarised the state of fisheries and outlined options for future management, including the quota management system. It was discussed at public meetings held throughout New Zealand during September 1983. The Tribunal noted that the meetings were widely publicised and attended by commercial fishers, other interested groups and the general public.[9] A proposal for a QMS for the inshore fishery was put forward by government in 1984. This too was widely discussed, initially with fishing industry leaders, and later at a second series of public meetings during late 1984 and early 1985.

The events occurring in the world of fishing could not be divorced from what was occurring in the realm of New Zealand politics. In July 1984, Prime Minister Rob Muldoon called a snap election. National was swept from power in a landslide vote bringing in the Labour government led by David Lange. The new government claimed that Muldoon had left the country virtually bankrupt and drastic measures had to be taken. State owned businesses would be sold and a market economy embracing the 'trickle-down theory' would be embraced. Entrepreneurs would create wealth through new business, assisted by lowered tax rates and fewer regulations, thus stimulating employment and other businesses and creating a better economy for everyone. Despite such neo-liberal economic policies, the new government was apparently empathetic and principled in its policies towards the Treaty of Waitangi and Māori claims against the Crown. In 1986 the judicial tide turned in favour of Māori, who have since enjoyed a number of successes in the courts.

The Fisheries Act 1983 was amended in 1986 (Fisheries Amendment Act 1986) to introduce the full QMS to New Zealand. The QMS put limits on the Total Allowable Catch (TAC) for each quota management area of each fish species included in the system (though not all commercial fish species were included at first). The TAC was then divided among those deemed to have rights to it – that is, those holding existing fishing permits. The amount allocated to each fisher, the individual transferrable quota (ITQ), was able to be sold or leased. The government in turn imposed resource rentals on the quota holders.

However, the element of 'protection' of Māori fisheries remained in the Fisheries Act [Section 88(2)]. The Law Commission pointed out that for more than a century, the Crown had declined to recognise any exclusive right that Māori might have to their fisheries, on the familiar grounds that there existed a common right of everyone to fish below the high water mark, subject to various regulations. However, the 1986 legislation established exclusive commercial fishing rights and gave these to a select few fishers. The Law Commission described this as:

> fundamental and far-reaching change, the essence of which is a shift from the previous effort-oriented controls (closed seasons, size of nets etc) to what was seen as a much more efficient and effective extraction-oriented control.[10]

The QMS in New Zealand

Doug Kidd recalled the introduction of the full QMS to the New Zealand Parliament in 1986 and the yet another round of consultation meetings with fishers throughout the country when the proposals were put forward in what was dubbed the 'Blue Book':

> an amendment to the *Fisheries Act 1983* . . . was going to update the penalty clauses. This is 1986. Most people didn't take any notice of the fisheries bill but even less people took notice of this one because it was boring as hell and only a few pages. It was going through in slow time because . . . what was the hurry. Made very little change – just needed to update penalties. Then, one day, one day, a SOP [supplementary order paper from the Minister of Fisheries] drops on the table in the house to attach to this tiny innocent uncontroversial Bill, it's the whole quota management system, hundreds of pages. I have to go back and say McIntyre started it and Moyle was the minister [of Fisheries] at that time. Anyway it got to the point where it couldn't be ignored. Inshore fisheries was in a state of ruin. They were subsidising the building of trawlers and the like. It was happening all around the world. I remember going to the first meeting over what was called the 'blue book' [government proposals for a quota management system for New Zealand fisheries], in Kaikoura, my electorate . . . It was going to lead to part-time fishers' permits being cancelled – pretty rough stuff. It was also proposed that there would be a 60% reduction in catch effort for most in-shore species that most New Zealanders knew the names of [Blue Cod, Snapper, Tarakihi] . . . And trying to get people to understand the only way to manage fisheries was biologically, biological data, we had to get it down to simple terms. I would say to people: 'The number one policy in fisheries is very plain; fish need to have a successful sex life, and if they don't, it's all over.' I knew that things were developing because they were taking the blue book around the coast and I went to the first one [consultation meeting] in Kaikoura. Needless to say Tipene O'Regan was there wasn't he and of course he's related to me.* He was perfectly able to speak (I only acquired the skill in a very modest way) in any crowd he was in. It was like you were in the . . . theatre. So, off he would go. He gave everyone their . . . pedigree, about their fisheries and most of the dire things he predicted came true.
>
> As I said, sooner or later in comes this supplementary order paper, nearly every piece of legislation has one or more, but the whole fisheries legislation, the whole fisheries law on the end of this poor innocuous little penalty update bill. Well my speech was priceless, constitutional outrage. Geoffrey Palmer never went to higher flights. Well Māori went clean off, understandably, and they said the effect of this was that Māori fishing rights were being destroyed and they could never get justice from the Crown once it was all allocated, privatised.[11]

---

*The relationship that developed between Doug Kidd and Tipene O'Regan, who were distantly related, was to be a critical factor in the Māori fisheries story and is highlighted in Chapter 8.

In its Muriwhenua report, the Waitangi Tribunal found that the QMS:

> is in fundamental conflict with the Treaty's principles and terms, apportioning to non-Maori the full, exclusive and undisturbed possession of the property in fishing that to Maori was guaranteed. Like the right envisaged by the Treaty, the quota right of fishermen is held in perpetuity but may be sold, or some other agreement or arrangement may be settled upon.
>
> The system has not only excluded the Muriwhenua people, but it has placed real difficulties upon new fishermen getting in.
>
> The Quota Management System may be contrary to the Treaty, but only as it is presently arranged. As we have said, the Treaty contemplated that new agreements may be made. There are many good features to commend the quota system, and if agreement can be reached, it appears that Maori interests could be accommodated within it.[12]

So the scene was set for a showdown. During the hearings of the Waitangi Tribunal in Muriwhenua, the Muriwhenua people called on the Tribunal to stop the QMS. The Tribunal in turn called on the government to do so, but this was rejected and the Ministry of Agriculture and Fisheries continued to allocate quota to fishers. This is where the court case in which Tom Te Weehi was claiming his right as Māori to fish became highly relevant.

In 1984, Tom Te Weehi had been caught with a number of undersized paua which he had collected for his personal consumption. He was consequently charged with breaching the Fisheries Act. He defended this charge by stating that he was collecting the pāua as part of his customary fishing rights under Section 88(2) of the Fisheries Act 1983: 'Nothing in this Act shall affect any Maori fishing rights'. Initially Te Weehi was found guilty, but he appealed to the High Court.[13] Evidence was given by a Ngāi Tahu elder and a member of the New Zealand Māori Council, William Karetai and Billy Nepia. Karetai said that the Ngāi Tahu people were at that time involved in discussions with the government over fishing rights, that they had always claimed fishing rights over their South Island coastline, and that the area from which Te Weehi had been gathering pāua was claimed to be a Ngāi Tahu fishing ground.

In August 1986, Justice Neil Williamson delivered a long and considered judgement.[14] He summarised a number of previous unsuccessful cases involving Māori challenges to New Zealand law regarding Māori fishing rights, and he looked in detail at the words of the current Fisheries Act that appeared to protect Māori fishing rights. He concluded that Te Weehi was exercising a customary Māori fishing right, and therefore other parts of the Act did not affect his right to take the pāua and he did not commit an offence. The appeal to the High Court was granted and the conviction was quashed.

This was the first time that Māori customary fishing rights were successfully used as a legal defence, giving all Māori the expectation that further challenges would be successful. The Crown did not appeal this decision. Sir Tipene O'Regan recalls that Ngāi Tahu were very concerned, since the decision could have led to 'plunder' by Māori exercising what they saw were their rights to the 'tribe's traditional resources'. As a holding measure, the Ngāi Tahu Maori Trust Board imposed a rāhui (traditional prohibition) 'aimed at holding the status quo under the Fisheries Act'.[15]

With the introduction of the QMS, in the midst of the Tribunal's Muriwhenua hearings and the High Court's delivery of a significant new decision recognising Māori customary fishing rights, this story takes a dramatic turn. There was continuing, simmering discontent among the hundreds of part-time Māori fishermen from around the country who had been excluded from the QMS and therefore from the fishing industry. Armed with the Te Weehi decision and the preliminary findings from the Tribunal's Muriwhenua consideration of fishing rights, the Māori Council and Muriwhenua leaders would take an unprecedented step of their own.

ENDNOTES

1. Waitangi Tribunal, *Ngai Tahu Sea Fisheries Report*. Department of Justice, 1992. pp. 217–19.

2. Doug Kidd, interview with Brian Bargh, 2014.

3. Peter H. Pearse, 'Fisheries Management Regimes for the Future.' XVIIIth Annual Conference of The European Association Of Fisheries Economists, Reykjavik, Iceland, 2007. pp. 1–7. www.univ-brest.fr/gdr-amure/eafe/eafe_conf/2007/peter_pearse_eafe2007.pdf

4. Waitangi Tribunal, *Muriwhenua Fishing Report*. Department of Justice, 1988. p. xviii.

5. Waitangi Tribunal, 1992. pp. 217–19.

6. Waitangi Tribunal, 1988. p. xviii.

7. Ibid.

8. Ibid.

9. Waitangi Tribunal, 1992. p. 222.

10. Law Commission, *The Treaty of Waitangi and Maori Fisheries*. Prelim. Paper No. 9, 1989. p. 18.

11. Doug Kidd, interview with Brian Bargh, 2014.

12. Waitangi Tribunal, 1988. p. xx.

13. http://www.fish.govt.nz/_indigenous_rights.pdf

14. Ministry of Justice, High Court Judgement M662/85.

15. Tipene O'Regan, 'Treaty Settlements, Fisheries and the Restoration of Rights'. Thomas Cawthron Memorial Lecture (unpublished), Nelson, August 1999. p. 7.

FIRST BITES:
Recovering Māori fishing rights

Before the Waitangi Tribunal first ventured to Northland in October 1986, Māori in the north had not really experienced the Tribunal, despite Sir Graham Latimer having been a member of it. The Tribunal had heard a number of claims from tribes south of Auckland, but now Māori of the North were able to see how the Waitangi Tribunal worked. From those observations they gained confidence that this was a way to air their grievances and trust that something could be done about them. When the Mangonui Sewerage claim hearings began, the claimants, Ngāti Kahu (including Sir Graham Latimer, Rev. Maori Marsden and McCully Matiu), were represented by Shonagh Kenderdine. Shane Jones and Mira Szaszy were appointed 'interpreters'.[1] The Tribunal itself included Eddie Durie (as Chair), Bishop Manu Bennett, Monita Delamere, E. D. (Ned) Nathan, Prof. Keith Sorrenson and Georgina Te Heuheu. The Mangonui Tribunal was sitting some six months ahead of the Muriwhenua Tribunal, although both reports were released in 1988. Membership of the Muriwhenua Tribunal was almost the same as the Mangonui Tribunal, except for Ned Nathan, who passed away in 1987. W. M. (Bill) Wilson came onto the Muriwhenua Tribunal in his stead.[2]

The Mangonui Tribunal did not make a recommendation, although it strongly suggested, among other things, that in siting ponds to treat sewage from the town of Taipa on Ngāti Kahu land, not enough weight had been given to Ngāti Kahu values.[3] This Tribunal report was not yet available when, in December 1986, the first hearing of the Muriwhenua claim got underway at Te Reo Mihi marae, Te Hāpua. That initial hearing lasted four days and was 'disrupted' by claimants drawing attention to the passage of the State-Owned Enterprises Bill in Parliament. They asked the Tribunal to intervene, because land held by government was to be transferred to the new state-owned businesses. This would, they claimed, put it beyond reach as redress for their claims. The full Tribunal agreed, and immediately wrote to the Minister of Māori Affairs, Koro Wetere, asking that lands in the Muriwhenua claimant rohe (territory) not be transferred to state-owned businesses. They also pointed out the implications of the new legislation for the forty or so claims by Māori up and down the country. The memorandum ended with these words:

> These are matters on which you may wish to confer further with the Minister of Finance [at the time Roger Douglas] and the Chairman of the Cabinet Committee on State Corporations. We trust you will consider them too in your capacity as Minister of Lands, Forests, and Maori Affairs. They constitute the message to you, at the head of the fish, from the Iwi at its tail.[4]

A further disruption to proceedings occurred when, as Shane Jones recalls, Captain Jacques Cousteau, the famous French explorer, conservationist, filmmaker and scientist, arrived at Te Hāpua. The Rev. Maori Marsden informed the Tribunal that the hearing would be suspended for the day, much to the displeasure of Tribunal member Monita Delamere, while Muriwhenua welcomed Cousteau. They also arranged for him to film some of the

The Waitangi Tribunal began its hearings into the Muriwhenua claims in December 1986 with a visit to Te Rerenga Wairua. Here, Tribunal members and Muriwhenua leaders hold a service to mark the start of the claim hearing in 1986. From the left: Tangi Ihaka (1st); Maarire Goodall (5th); Raupo Brown (6th); Lewis (Nuk) Brown (7th); Tribunal Chairman Eddie Durie (9th); Matiu Rata (partly obscured 10th); Waerete Norman (11th); Paul Harman (holding umbrella); Monita Delamere (Ringatu Church) and Bishop Manuhuia Bennett (both under umbrella); Manawa Aperehama (partly hidden by umbrella,16th); Hone Brown (Hone Rehu), 17th; Rapine (Robin) Aperehama, Ratana Church and kaumātua of Ngāti Kuri 18th; Bill Brown (21st); Rapata Rapata (Kut), 23rd.

Adam Gifford

hearing and the people involved. This programme was later played on New Zealand television, with the aim of leading to a greater understanding of the Tribunal's role and operation.[5]

David Baragwanath QC was appointed counsel for the Muriwhenua tribes in their fisheries claim. Shane Jones, who, with Mira Szaszy, was appointed 'translator' to the Tribunal, said that initially the Muriwhenua people were somewhat wary of Baragwanath. But some of the kaumātua recalled his father, the Rev. Owen Baragwanath OBE, former moderator of the Presbyterian Church in New Zealand, and that paved the way for David's acceptance.[6] Baragwanath was later joined by Sian Elias; both were to lead distinguished careers in law (Sian Elias is currently Chief Justice of New Zealand) and both were knighted. Strangely, to the Muriwhenua people, Shonagh Kenderdine was the lawyer for the Crown in the Muriwhenua fisheries claim, after having been the claimants' lawyer in the Mangonui Sewerage case.[7]

On 30 September 1987 the Waitangi Tribunal, while conducting the last of the Muriwhenua hearings, sent an urgent memorandum to the Minister of Fisheries. In it the Tribunal made some strong 'observations' about the fisheries of the Muriwhenua tribes, because of its concern that the government pressing ahead with the allocation of quota would seriously compromise its findings and the government's ability to provide redress for Muriwhenua. These concerns had first been raised in the first Muriwhenua hearing when Harry Mikaere from Hauraki attended and briefed Muriwhenua claimants and counsel. Claimants then alerted the Tribunal. However, at that early stage the Tribunal was powerless to intervene, although they did convey their concerns to the Director-General of Fisheries, asking him to cease allocating fish quota until the Tribunal could consider the matter. The Director-General's reply claimed that the quota system was already proceeding.

In the September 1987 memo, the Tribunal stated that Muriwhenua indeed 'owned' the area of sea off their territory and that the Crown did not have the right to allocate fisheries in this area; rather they should negotiate with Muriwhenua in order to get their agreement to the exploitation of the fishery. The memorandum had obvious implications for fisheries around the whole of New Zealand.

That afternoon, Muriwhenua (led by former Labour Government Minister of Māori Affairs and Lands, Matiu Rata) and the New Zealand Māori Council (led by Sir Graham Latimer) made an urgent appeal to the High Court, with Justice Greig presiding. The Court agreed to order a halt to the government's implementation of the QMS for squid and jack mackerel in waters around Northland. The fishing season began the same day. The claimants had made this last-minute oral application for an injunction because they said that their claim to the Waitangi Tribunal would be prejudiced if the QMS was continued. Matiu Rata, chairman of the Muriwhenua Incorporation, said the injunction was sought after talks with the government had been unsuccessful.[8] Justice Greig's reasons for his orders were that:

> There was a strong case that before 1840 Maori had highly developed and controlled fishery over the whole coast of New Zealand. This control ('perhaps rangatiratanga') was divided along hapu and tribal boundaries and the fisheries were sustaining, recreational, ceremonial and commercial. The judge also found that Maori had not waived or given away their fishery rights but that since 1840 'there has been a great diminution and a restriction in Maori fishing through circumstances, to use a neutral word, which in the end have limited the exercise of those rights'.[9]

Significantly, the judge ruled, Māori fishing rights had not been taken away.

The Ngāi Tahu Maori Trust Board also took action to prevent the government implementing the quota system for squid and jack mackerel in South Island waters by filing another application with the High Court. Ngāi Tahu felt that the tribe's Waitangi Tribunal claim would be prejudiced if the quota system was allowed to go ahead. Henare Tau of Ngāi Tahu said he encouraged all other tribes to take similar action.[10]

Meanwhile the commercial fishing industry was in turmoil. The industry was already upset by the QMS and the prospect of having to pay resource rentals for their quota. The Fishing Industry Board was concerned that some areas around the coast would be excluded from the QMS because of the High Court injunction, and were worried that the exclusion of two fish species because of the injunction would badly affect commercial fishing.[11] Incidents were occurring that served to heighten tension. For example, two Northland commercial fishermen claimed they had been threatened by a group of Māori. The Northland branch of the Federation of Commercial Fishermen chairman, Bob Martin, said the action was due to inflammatory statements made by the Māori Council on fishing rights. However, Māori Council chairman Sir Graham Latimer said that the commercial fishermen were trawling too close to the shore and that resulted in local Māori confronting them. He said the claim by Martin was 'absolute rubbish'.[12]

Negotiations begin

The High Court interim injunction meant that the government could not continue implementing the QMS. The government moved rapidly to set up a joint Crown–Māori working party, to be funded by the Crown. Headed by Justice J. H. Wallace, the Crown team included John Chetwin from Treasury, Fred Baird from the fishing industry, and the former managing director of Dominion Breweries, Ross Sayers (also recently appointed chair of the Crown-owned Railways Corporation). Sir Graham Latimer, Matiu Rata, Tipene O'Regan and Denese Henare represented Māori, with Denese Henare representing Bob Mahuta and the Tainui Maori Trust Board. The working party was due to report by June 1988 on ways of giving effect to Māori fishing rights.

In order to cause less disruption to the fishing industry, in December 1987 the High Court issued two interim orders confirming that the QMS could continue for the current fishing season for an additional two fish species. Justice Greig noted that 'Maori people do not by one whit step back from their stance that the Crown has no power to license or permit fishing within Maori fisheries'. Meanwhile a delegation of northern commercial fishermen had met with the Minister of Fisheries, Colin Moyle, to obtain assurances that they would be consulted over proposed changes to Māori fishing rights.[13]

The Muriwhenua Fisheries Report

With the joint working group due to deliver its report by 30 June 1988, in May the Waitangi Tribunal issued its full Muriwhenua Fishing Report recommending 'very substantial relief' for claimants. The Tribunal said that blatant and serious breaches of the Treaty guarantee had occurred and Muriwhenua had lost a valuable industry [fishing]. The Tribunal recommended that Muriwhenua Māori be given quota under the QMS; in return they would essentially suspend or give up their Treaty fishery rights for 20 years, and the Waitangi Tribunal would be barred from further inquiry into those rights for the same period.[14] This release only added to the frenzy.

Opposition Māori Affairs spokesman Winston Peters said the implications of the report were enormous.[15] The next day, fishing industry leaders met the government members of the joint Crown–

OPPOSITE PAGE

Dame Whina Cooper being pushed by Eddie Kawiti talking to Rev. Maori Marsden, during the Ngāti Whātua claim hearings, Auckland, May 1985. Maori Marsden was a tohunga, scholar, writer, healer, minister and philosopher in the latter part of the twentieth century; like Whina, he was from the Far North. He was a significant force in Māori protests from the 1960s until his passing in 1993.

Gil Hanly

ABOVE

Matiu Rata, at home c. 1980. He established the Mana Motuhake Party when he resigned from the Labour Party in 1979, following his rejection of the policies of the Labour Party. He wanted Māori to have an independent voice in Parliament, but neither he nor any of the new party candidates were ever elected. Here Matiu presides over a party meeting with another founding member, Amster Reedy. Matiu devoted much of his time after leaving Parliament to leading the Muriwhenua claim and advancing the campaign for Māori fisheries rights.

Gil Hanly

Waitangi Tribunal members Sir Graham Latimer (right) followed by Chair Eddie Durie and Paul Temm QC (partly obscured) move forward to take their place at the Ngāti Whātua o Ōrākei marae hearing, Auckland, May 1985.

Gil Hanly

Māori working party. Vice-President Peter Talley called the Waitangi Tribunal findings 'legalised apartheid'. The fishermen believed all fisheries should be vested in the Crown. The fishing industry leaders were upset and worried at what the future might bring. Members of Parliament John Banks and John Carter were highly critical. Banks reportedly called the report's findings legalised racism and Carter 'voiced fears of bloodshed between Maori and Pakeha'.[16] Minister of Justice Geoffrey Palmer and Prime Minister David Lange moved to allay fears.[17] Parliament held a special debate on the Muriwhenua Fisheries Report. Opposition leader Jim Bolger called for the debate because, he said, there was growing disquiet among Māori and Pākehā over the status of the Treaty of Waitangi and of government initiatives that led to heightened expectations among Māori. David Lange said that the government would act on the Muriwhenua report recommendations, but that the Crown's sovereign rights to manage fisheries would be preserved. He declined to assure Parliament that Māori would have the same rights as other New Zealanders with regard to fishing.[18]

Meanwhile, rumours were circulating as to what the joint working party on Māori fisheries was going to recommend. It was said that Crown members were suggesting a joint Crown–Māori fisheries corporation where profits would be split between Māori and the Crown. Māori negotiators refused to comment on the proposal. However, Deputy Prime Minister Geoffrey Palmer said that the proposal had merit. Justice Wallace, convenor of the joint working party, said he would not comment as negotiations were still continuing.

Jim Bolger called for 'time out' on the work of the Waitangi Tribunal until New Zealanders could see the way ahead. The Minister of Fisheries, Colin Moyle, said the government had no intention of abandoning the QMS and pointed out that the Waitangi Tribunal had praised the system. He also said the government would restore the Treaty rights of Māori 'as best we can'. Winston Peters said the Tribunal findings in the Muriwhenua report were a recipe for social disaster.[19] At the same time, The Dominion newspaper reported that the Takitimu District Māori Council and Tainui Maori Trust Board were claiming ownership rights to fisheries in their coastal waters. It also reported strong opposition to the Muriwhenua report from commercial fishermen.[20] The fishing industry was in an unsettled state and the public debate was raging, with comment in newspapers and other media ranging from informed to racist and ignorant.

On 24 June 1988, six days before they were due to report, the Māori fisheries negotiators from the joint Crown–Māori working party met with about 150 Māori representatives and the four Māori Members of Parliament. Maanu Paul, chair of the Waiariki District Māori Council, said that the meeting was the first time Māori had discussed among themselves the various options for giving effect to the findings in the Waitangi Tribunal's Muriwhenua report and the matters that had emerged from the joint working party. Māori negotiators believed that Māori and the Crown should share fisheries equally out to a 200 mile limit from the coast. The negotiators said that the Crown proposal of a corporation with 20 percent Māori shareholding, to control fisheries and receive income from rental of fish quota, was not acceptable.[21] After the meeting, Sir Graham Latimer announced that the Māori fisheries negotiators (himself, Denese Henare, Tipene O'Regan and Matiu Rata) had received a mandate from the national fisheries hui for their proposal. He said some representatives had called for 100 percent ownership by Māori out to 12 miles from the coast, but consensus was for a 50 percent share. It was proposed that the 50 percent share be transferred to Māori over 16 years in order to avoid disrupting the fishing industry. Tipene O'Regan said that going back to court was a possibility, but that would create political and social stress; in any case, the two parties would still have to negotiate a solution.[22]

While the Māori fisheries debate was in public focus it was not the 'only show in town'. Tainui Māori were faced with the Crown wanting to sell Coalcorp, the state-owned coal mining company. Tainui, who were also deeply involved in the fisheries dispute, were claiming land and the coal under that land in their Waikato area. The Tainui Maori Trust Board issued a statement hinting to government that it should not fall into the same trap as it had done with fisheries — selling fish quota without determining who owned that quota. It needed to establish who owned the coal before selling mining rights to it. The tribe said they believed that they owned the coal.[23]

The Waitangi Tribunal was also in the midst of its hearings into the Ngāi Tahu claim. To add fuel to the fire, the Chairman of the Ngai Tahu Maori Trust Board, Tipene O'Regan, told the Tribunal that Ngāi Tahu claimed all inshore and offshore fisheries within their territory of the South Island. He said that the Crown did not have the right to issue quota, licences or other forms of title to fish. However, the tribe was willing to accept an equal share in the management and control of the fisheries — 'this is a very considerable concession'.[24]

It seemed to some New Zealanders that matters were out of hand. For example, some commercial fishermen called for a referendum on the Treaty of Waitangi. Māori leaders responded by saying that public debate was already taking place and a referendum was not needed. Mana Motuhake political party leader Matiu Rata said he was infuriated by the constant attacks on the Treaty. He felt the fishing leaders should understand their obligations under the Treaty, rather than seek ways of undermining it: 'Every New Zealand citizen has the obligation to uphold the Treaty and I don't think there is any future in looking for new ways to welch on it.'[25] Five days later, the fishing industry's Bob Martin placed 20 large

advertisements in newspapers throughout New Zealand calling for a national referendum on the Treaty of Waitangi. These ads, headed, 'Is New Zealand still a democracy?', focussed on the Māori fisheries claims and sought donations to help stop what he saw as a 'takeover of the assets of this country'. In a rather hysterical tone, Martin said he wanted changes to the way racial issues were handled before the country was destroyed.[26]

The joint Crown–Māori Working Party reports

The joint Crown–Māori working party on Māori fisheries could not agree. It produced separate reports, which were presented on 30 June 1988. Although details had already been leaked to the news media, the main feature of the Crown's report was the suggestion of a Fisheries Corporation controlled jointly by Māori and the Crown. The corporation would be responsible for all aspects of fisheries management. Rights to fish in New Zealand would be vested in it and be allocated through the quota management system. The proposal would settle Māori fisheries matters, excluding freshwater fisheries. A training programme for Māori people to get involved in the fishing industry was also proposed.

The major disagreement with the Māori negotiators was over the Māori share of the inshore fisheries. Matiu Rata said that he felt the differences could be worked out. Māori claimed 50 percent of the fishery, but the Crown was offering only 29 percent. According to a report in *The Dominion* newspaper, fishing industry representatives had rejected both proposals.[27]

The Crown–Māori working party was then disbanded and Richard Prebble, associate Minister of Finance, took charge of reaching agreement with Māori. A group of Māori leaders (Sir Henare Ngata, Sir James Henare, Sir Ralph Love and Sir John Bennett, together with Bruce Gregory, MP for Northern Māori, and the Māori Fisheries negotiators, met with Labour government ministers Richard Prebble, Geoffrey Palmer and Colin Moyle to discuss the Crown–Māori working party reports, and agreed to release them to the public.[28] While the Māori negotiators had considerable respect for Geoffrey Palmer (who was then deputy Prime Minister) and David Caygill (finance minister – also involved in the negotiations), they were wary of Richard Prebble's tough and rather flippant negotiating style. Sir Graham Latimer recalls:

> You see he did his calculations exactly like this . . . while we were negotiating, they [the Crown including Prebble] kept on saying to us, how much fish do you want, and my story was we wanted the lot. And then they come around to the story of how much fish can you fish? And that's when we made the mistake because Tipene O'Regan said Ngai Tahu can fish 6,500 tons a year and Prebble quickly said, 'Well 6,500 tons for Ngai Tahu and 6,500 tons for the rest of the tribes and 3,000 tons of fish for commercial.' That brought us up to 16,000 tons which was 10 percent of the fish. So it was two and a half times 4 percent . . . supposed to be over 10 years, then we got him back on to the permanent allocation [they eventually agreed on four years to achieve a 10 percent total of quota to be given to Māori] and that's how it ended up.[29]

Tipene O'Regan recalls the negotiation slightly differently:

> Prebble said to me 'how much fish can Māori catch in a season?' Overnight I came up with a figure

OPPOSITE PAGE

David Lange and Geoffrey Palmer leaving the Labour caucus after the election of the Cabinet, August 1984. As a former law lecturer, Palmer maintained a principled stance throughout his government's dealings with various Māori Treaty grievances, including the Māori fisheries negotiations

ATL: EP/1984/3426/3-F. Ross Giblin, Evening Post *Collection*

which equalled 10 percent of the deep sea and inshore quota available which formed the basis of the interim settlement.[30]

The Crown seemed set on the notion that Māori had to fish their allocated quota, and this notion carried on into the agreement.

Three days later Richard Prebble declared that progress had been made on a number of points, and the government would consult with the fishing industry over the next week. Matiu Rata was upbeat about the discussions, saying that in a very short time they were close to 'one of the most historical decisions New Zealand is likely to make this century'.[31] The joint reports of the Crown–Māori working group were released the day before, due to the public interest. Prebble said that the government had not adopted any of the working group recommendations, and Ministers of the Crown were now working directly with Māori to try to reach agreement before the start of the next fishing season on 1 October 1988.[32]

Progress toward a solution seemed to be rapid, making fishing industry leaders very nervous. Richard Prebble assured them that whatever deal was reached over Māori fishing rights, it would not be at their expense. He was not able to agree to include industry representatives in the current negotiations because Māori did not agree.[33] A few days later, the Chairman of the Māori Council's fisheries sub-committee, Maanu Paul, said the fishing industry had no place in talks between the Crown and Māori. He complained in a letter to Prebble of the undue pressure the fishing industry was putting on the Crown and accused them of stirring up racial tension. This was a matter for Treaty partners, Paul said, adding that in the spirit of sharing between partners, Māori were prepared to offer the Crown 50 percent of the commercial fisheries. Pressure on the Crown came also from a large fishing company, Fletcher Fisheries, who said they had reconsidered going ahead with expansion of their fishing fleet due to uncertainty over resource rentals and Māori fishing rights. But Richard Prebble hit back at Fletcher, saying he doubted that such big decisions over expansion were made based on 'such ill-researched nonsense'. He said no-one was wanting to destroy the fishing industry.[34]

A July meeting of commercial fishermen in Wellington was told by their leaders that they needed to act fast to oppose Māori fisheries moves by government. The meeting also demanded that quota taken by the Crown to satisfy obligations to Māori be paid for at fair market prices. Resource rentals were also discussed.[35] The fishers were still suspicious that fish quota might be acquired compulsorily, though they probably would not have seen the irony in this. Letters to newspaper editors about this time were full of scorn over the fact that Māori were negotiating with the Crown over their fishing rights. There was continuing uncertainty, particularly from commercial fishermen, about what the outcome might be. Some suggested the Treaty of Waitangi be done away with. Others claimed the Treaty did not mention 'fish'.[36] On the other hand, a letter was circulated from a group calling itself the Māori Alliance to Members of Parliament calling for support for the Māori fisheries negotiators, and for Māori to grant the government a 50 percent interest in fisheries. The letter stated that the 50 percent offer was in the best interests of the nation, given that Māori had never ceded ownership of their fisheries.[37]

Agreement on Māori Fisheries is achieved

On 22 September 1988, fisheries Minister Colin Moyle introduced into Parliament a Māori Fisheries Bill. In its long title, this Bill spelt out certain terms of agreement reached between Māori and the Crown, which, *inter alia,* provided over a 20 year period for the Crown to acquire and then hand to Māori, in each year from 1989 to 2008, quota representing 2.5 percent of the total allowable catch, thus giving Māori a 50 percent interest in sea fisheries. A grant of $10 million spread over five years was also to be made to help Māori tribes to engage in the business of fishing. In return, Māori rights to commercial fisheries would be extinguished.

The 1988 Bill was referred to a Select Committee, where the fishing industry strongly contested it. Māori also criticised several proposals in the Bill. With the threat of loss of their fisheries looming, further proceedings were filed in the High Court, and there was strong debate in the media. Almost all fishing tribes brought actions alleging trespass, breach of fiduciary duty and negligence. This time it was political considerations that forced the Crown back into negotiations. The Crown could have continued with the legislation and ousted the court's jurisdiction. However, this option was seen as unconstitutional and risky.

The Select Committee considering the original Maori Fisheries Bill incorporated some changes, which met with the approval of the negotiating parties and resulted in the passage of the Maori Fisheries Act 1989 on 20 December 1989. This was an interim settlement that consisted of $10 million cash to be transferred to a Māori Fisheries Commission; the establishment of Aotearoa Fisheries Limited; and 10 percent of the commercial fishing quota to be managed by the Commission, on behalf of Māori. Where quota could not be obtained by the Crown, a cash equivalent would be paid to the Māori Fisheries Commission. These assets, together with further quota and other assets acquired in the market, were later to become known as the Pre-Settlement Assets.

It was then agreed by those parties before the courts, namely Māori, the Crown, and the New Zealand fishing industry, to adjourn all proceedings *sine die* to give the new Act a chance to establish a transfer mechanism by which the Crown would issue quota to iwi, and establish taiapure fishery reserves in recognition of Māori customary fishing rights. The transition period was to expire on 31 October 1992, in approximately three years' time.[38]

The new Maori Fisheries Act 1989 was an interim measure. Under that Act, Māori were to be given quota in four equal tranches of 2.5 percent over the period from April 1990 to October 1992. It was made clear to the Tribunal by Ngāi Tahu claimants that Māori did not accept the 10 percent quota as representing a full settlement of Māori fisheries claims. The object of the Act was to facilitate Māori entry into the business and activity of fishing and to allow a settling down period. It was the Ngai Tahu Sea Fisheries Tribunal Report that described the Act as a 'breakthrough towards Crown recognition of Māori Treaty rights'.[39] Additionally, rock lobster was brought into the QMS (with 10 percent reserved for Māori) and the Act provided for taiapure (non-exclusive local fisheries) to be established.

In terms of recognition of Māori rights, the most important measure was the establishment of the Māori Fisheries Commission, which was to consist of seven members appointed on the advice of the Minister of Māori Affairs. Among its principal functions was to facilitate and foster Māori into the business of fishing. It was required to form and be sole shareholder in a public company called Aotearoa Fisheries Limited, and to transfer to that company at least 50 percent of all quota and money transferred to the Commission by the Crown under the Act. All quota and money so transferred was to be applied in paying up in full the shares in the company to be issued to the commission.[40]

By a later amendment to the Act, the Commission was required to consult with Māori tribes who had interests in fishing (which was, of course, virtually all tribes). The consultation was mainly to reach agreement on the best way of transferring the quota from the Commission to those entitled, and to determine their entitlement. In the meantime, the Commission was to operate closely with Aotearoa Fisheries to manage the quota so that its value was maintained and enhanced.

After ignoring Māori Treaty rights to fisheries for almost 140 years, the Crown was not only forced to recognise and give effect to those rights, but was charged with compensating Māori for their long suffering, the trampling of their mana and their exclusion from the development of fisheries that would no doubt have seen them becoming successful leaders in that industry.

The year 1990 was one of new beginnings for Māori fisheries. A Māori fisheries commission was to be established, a new Māori fishing company to be created and Māori generally to be launched into the business of fishing. The task appeared huge.

ENDNOTES

1. Shane Jones, pers. comm., March 2015.
2. Waitangi Tribunal, *Mangonui Sewerage Report*. Department of Justice, 1988. p. 61.
3. Waitangi Tribunal, *Mangonui Sewerage Report*, 1988. p. 66.
4. Waitangi Tribunal, *Muriwhenua Fishing Report*. Department of Justice, 1988. p. 291.
5. Shane Jones, pers. comm., March 2015.
6. Ibid.
7. Ibid.
8. *Dominion*, 1 October, 1987. p. 1.
9. Waitangi Tribunal, *Ngai Tahu Sea Fisheries Report*. Department of Justice, 1992. p. 231.
10. *Dominion*, 23 October, 1987. p. 3.
11. *Dominion*, 2 October, 1987.
12. *Dominion*, 7 December, 1987. p. 1.
13. *Dominion*, 12 December, 1987. p. 2.
14. Waitangi Tribunal, 1992. p. 233.
15. *Dominion*, 13 June 1988. p. 1.
16. *Dominion*, 14 June 1988. p. 1.
17. Ibid.
18. *Dominion*, 15 June 1988. p. 1.
19. *Dominion*, 16 June 1988. p. 1.
20. *Dominion*, 17 June 1988. p. 10.
21. *Dominion*, 25 June 1988. p. 6.

22. *Dominion*, 27 June 1988. p. 1.
23. *Dominion*, 25 June 1988.
24. *Dominion,* 28 June 1988. p. 9.
25. *Dominion,* 23 June 1988. p. 2.
26. *Dominion,* 28 June 1988. p. 9.
27. *Dominion*, 28 June 1988. p. 7; 30 June 1988. p. 7.
28. *Dominion*, 6 July 1988. p. 2.
29. Graham Latimer, interview with Adam Gifford, 1995.
30. Tipene O'Regan, pers. comm., September 2014.
31. *Dominion,* 8 July 1988. p. 2.
32. Ibid.
33. *Dominion*, 12 July 1988. p. 1.
34. *Dominion,* 18 July 1988. p. 2.
35. *Evening Post*, 21 July 1988. pp. 1, 2.
36. See, for example, 'Letters to the Editor', *Dominion*, 25 July 1988. p. 15.
37. *Evening Post,* 13 August 1988.
38. See fuller account in Waitangi Tribunal, 1992. p. 234.
39. Waitangi Tribunal, 1992. p. 237.
40. Ibid, p. 238.

ALL ABOARD THE TRAWLER:
Consolidating the gains

At the start of 1990, the eyes of the Māori world in particular were upon the proposed Māori Fisheries Commission. Many of the tribes were eagerly awaiting the allocation of quota so that they could get into commercial fishing. However, the reality was to be somewhat different. There was a maze of bureaucratic procedure to wade through, a myriad of officialdom to crank into action and numerous critics who, it seemed, wished the new Māori venture to fail.

In February 1990 the Minister of Māori Affairs, Koro Wetere, appointed the first Māori Fisheries commissioners: Tipene O'Regan (chair), Dame Mira Szaszy, Whaimutu Dewes (one of the first Māori lawyers in the 1970s), Dr John Mitchell (a leader from the top of the South Island), Stephen Jennings (a Taranaki businessman), Nick Jarman (previously from the Fishing Industry Board) and Sir Graham Latimer (deputy chair). Discussions also began between the Māori fisheries negotiators and Crown fisheries representatives about the continued development of the QMS in a manner that met both conservation requirements and the principles of the Treaty of Waitangi. They also agreed that further court proceedings would be suspended, and that no further species would be introduced into the QMS until a satisfactory resolution of matters was reached.

In July 1990 the Hui-ā-Tau (annual meeting) of tribal fisheries representatives was convened by the Commission to discuss allocation of quota and other matters following the momentous events of the previous few months. The meeting agreed that quota being received from the Crown by the new Commission belonged to tribes, and would be allocated to them from October 1992. There was opposition to the formation of Aotearoa Fisheries Ltd (AFL), the company which would receive half the quota transferred to the Commission, because the iwi representatives saw that as taking away half the quota they could be using.[1] It was revealed that the Māori negotiators had only reluctantly acceded to AFL being part of the agreement, since they saw it as a Crown mechanism to prevent Māori tribes receiving and using all their allocated quota. Tipene O'Regan said of the interim settlement: 'It's not a final settlement but it's a chance to make a start and get structures into place ready for the rest of our 50 percent. The return of our property rights in fish is one thing. Developing the skill to maintain and hold them for our mokopuna is the next great challenge'.[2] Sir Graham Latimer was appointed chair of AFL in June 1990, but the first AFL board meeting was not held until September 1990. Apart from delivering a 'Statement of Intent' and company infrastructure, there was intense behind-the-scenes debate between Koro Wetere and his advisors, the Māori Fisheries Commissioners, and others over who the directors of AFL should be. Those finally appointed were: Sir Graham Latimer (Chair), Craig Ellison, Ken McDonald, Russell Armitage, Harry Mikaere, Annette Sykes and Rob Challinor. The initial assets of the company consisted of offshore quota.

Meanwhile, the Crown set about obtaining 10 percent of quota in all fish species already in the QMS, in order to fulfil its agreement with Māori. Fortunately, the Fletcher Fishing Company was available for purchase and owned both inshore and deepsea offshore quota. The Crown purchased the company's inshore quota and transferred this to the Māori Fisheries Commission as part settlement of the required 10 percent. The deepsea quota was sold to Sealord, Ngāi Tahu and others. However, northern tribes were not satisfied because quota for the fish species they were interested in (particularly snapper) was not available to the Crown, and Matiu Rata was threatening further legal action. Finance Minister David Caygill was particularly helpful to the Commission at this stage.[3] Sir Tipene O'Regan recalled that Caygill was:

> highly principled . . . also extraordinarily helpful in finding a solution to the problem [transferring the agreed 10 percent of fish quota to the new Commission] because the Crown did not have the quota to cover its obligation. Caygill brought in a model whereby it [the Crown] paid us the rental value of the missing quota in the season in which the quota was not delivered . . . And when they could find it they would hand the quota over to us but we wouldn't have to pay the rent back . . . on the basis that we had lost the opportunity to get that rent because of the time it had taken to get the quota. This mechanism was proposed by us, accepted by Caygill and implemented by us.[4]

National politics took the headlines during this beginning period for the Commission. Geoffrey Palmer had become Prime Minister after David Lange resigned in 1989; but over the next few months it was obvious to Labour Party supporters that Palmer did not have Lange's charisma and leadership abilities. In a desperate attempt to win the upcoming 1990 general election, the Labour Party dumped Palmer and installed Mike Moore as Prime Minister. It did not work. The years of turmoil wreaked by the Lange government and its neoliberal economic policies (state owned asset sales, imposition of a Goods and Services tax, 'trickle-down economics', and other reforms that were unpopular) saw party supporters desert. As well, the share market crash of 1987 had badly affected business and investor confidence and Labour was lucky to have won a second term of government.

On 27 October 1990, Labour lost the election and a National Party government, led by Jim Bolger, took control. Doug Kidd became Minister of Fisheries and Winston Peters was appointed Minister of Māori Affairs. Doug Graham became Minister of Justice and was made responsible for Treaty negotiations. These men, particularly Doug Kidd, were to make significant contributions to the story of Māori fisheries.

The development of the Māori fishing industry

In May 1991 Robin Hapi was appointed as the first general manager of the Māori Fisheries Commission.[5] His task of steering the commission toward its goals of protecting and growing its assets and getting Māori people into the fishing industry was difficult and demanding. The simmering argument between tribes with large coastlines and relatively smaller populations (such as Ngāi Tahu) and those with large populations but smaller coastlines (such as Ngā Puhi) heated up. There was also continuing resentment about the requirement for the Commission to transfer 50 percent of the quota and money received from the Crown to Aotearoa Fisheries Ltd. Some felt that all the quota should be allocated directly to tribes to manage themselves.

In June 1991 Sir Graham Latimer, speaking as chairman of AFL, said he opposed moves by some iwi to wind up AFL. He wanted to leave the quota vested in AFL as a pan-tribal venture after October 1992, rather than start allocating it to iwi. Sir Graham was reflecting on a number of recent consultative meetings with iwi around the country. He believed that once AFL started performing well, iwi would want to retain it. He admitted that when the Labour government had first proposed setting up AFL he had been

TOP

John Fernyhough hands Roger Douglas and Richard Prebble an Electricity Corporation cheque for $6.3 billion, for Crown electricity assets, Wellington, April 1988. A huge number of New Zealanders opposed the sale of state-owned assets and the Labour Party lost power in 1990, in part because the electorate became disillusioned with its free-market policies.

ATL: EP/1988/1693/20a-F. Ross Giblin, Evening Post

LEFT

Sir Tipene O'Regan (left) with Sir Graham Latimer announce a new decision by the Commission on allocation of fisheries assets, Wellington, 17 April 1997. These two outstanding leaders toiled long and hard to achieve the fisheries settlements.

Fairfax NZ: O'Regan Dave Haniford

The East Coast rock lobster (crayfish) fishery was closed in September 1993 to allow stocks to recover. When the fishery was re-opened fishers hauled record catches. Here the first crayfish into Moana Pacific's Gisborne plant after the moratorium are examined by LeRoy Pardoe (left) and Greg Griffiths, February 1994.

TOK

opposed to it.[6] However, at one consultative hui held in Dunedin, Ngāi Tahu representatives said that they wanted to take their allocation of quota and manage it themselves, rather than rely on it being managed by AFL.[7]

Meanwhile Māori tribes and groups were quickly seizing the initiative. Ngai Tahu Fisheries Ltd had recently been set up to manage the tribe's fishing quota and develop a fishing business. In the 1990–91 year they leased $1 million worth of quota from the Māori Fisheries Commission.[8] *Tangaroa*, the official newsletter of the Commission, also reported that six iwi and other Māori organisations were to form a joint venture company with the Skeggs fishing company. They used their quota asset to raise the funding. Another large fishing company, Moana Pacific Ltd, had been formed the year before with equal shares by Wilson Neill (an existing fishing company) and the Māori Development Corporation (formed to attract investment from various Māori tribes and organisations, and in turn to invest in productive enterprises).[9] Moana Pacific bought the inshore quota of Fletcher Fishing. This subsidiary company of the large and expanding Fletcher Challenge firm originally held 18 percent of all New Zealand fishing quota. Its deepsea assets and quota were sold to Carter Holt Harvey's subsidiary Sealord Products. Part of its remaining quota was sold to the Crown to be used to satisfy obligations to provide quota to the Māori Fisheries Commission, and its remaining inshore assets were sold to Moana Pacific. Both Moana Pacific and the new joint venture company formed with Skeggs were developing training schemes for Māori to get them into the business of fishing.[10]

The directors of AFL followed a policy of acquiring fishing assets as a way of raising the value of their assets held on behalf of iwi. They purchased a major shareholding in Moana Pacific Fisheries Ltd from Wilson Neill. This meant that Moana Pacific was now fully owned by Māori and was the largest inshore fishing company in New Zealand, with valuable quota in rock lobster, snapper and orange-roughy. The majority of its products were exported.[11]

ENDNOTES

1. *Tangaroa,* Treaty of Waitangi Fisheries Commission newsletter, No. 1, September, 1990. p. 3.
2. Ibid, p. 1.
3. Tipene O'Regan, pers. comm., 2014.
4. Tipene O'Regan, interview with Brian Bargh, 2015.
5. *Tangaroa*, No. 5, June 1991. p. 6.
6. Ibid, p. 3.
7. Ibid, p. 4.
8. Ibid, p. 8.
9. Ibid, p. 7.
10. Ibid, p. 7.
11. *Tangaroa*, No. 6, July 1991. pp. 1–2.

SETTING THE NETS:
Sealord negotiations

Following the passing of the Māori Fisheries Act in December 1989, the new Māori Fisheries Commission had begun work on considering how the assets it held in trust should be allocated to the recognised tribal fishing entities. Reaching agreement was extremely difficult, but by December that year the Commission, after consulting widely, had resolved to accept that the over-arching principles for allocation of quota to tribes would be based on mana whenua (the extent of a tribe's authority over its lands) and mana moana (a tribe's authority over its seas adjacent to those lands), because fishing rights were collective in nature. At the same time, the fishing industry was in a state of rapid change, with companies merging, being sold, looking for finance, buying quota and so forth. Māori-owned companies were mostly expanding and looking for opportunities.

In October 1991 Winston Peters had been sacked as Minister of Māori Affairs and replaced by Doug Kidd, who was also Minister of Fisheries. He had already been busy as well, meeting secretly with Tipene O'Regan in his office, usually on a Sunday evening, when they would discuss fishing and other things. The two had become trusted friends after being involved in earlier fisheries matters and also, in different ways, in the Taranaki 'think big' projects. This relationship was to become pivotal in the final settlement of the Māori Fisheries dispute.

January 1992 saw the beginning of a remarkable string of events, perfectly timed, which resulted in some remarkable Māori leaders landing an unexpected catch. Both Māori and the Crown – particularly Minister of Fisheries Doug Kidd – were well aware of the problem that would confront them at the end of 1992 when the 'interim' fisheries settlement period would end and a practical definition of Māori fisheries would then be thrashed out in the High Court. It was clear there would still need to be negotiations between the Crown and Māori to produce a settlement that satisfied Māori expectations, settled the wider fears and uncertainty in the fishing industry and was acceptable to the general public. Only then could any certainty be achieved. The Chair of the Commission, Tipene O'Regan, was maintaining that Māori expected a fair settlement to give them 50 percent of the total commercial catch. Doug Kidd announced in March 1992 that his talks with the Māori negotiators would begin again, in an effort to reach agreement and have the High Court injunction lifted from the QMS.[1]

Recalling his regular meetings with Tipene O'Regan during 1991 and 1992, Doug Kidd remembered:

> When I became Minister (of Fisheries) the date of proceedings was looming and we needed a resolution. So I started to talk to Tipene, usually at night, so there wouldn't be any witnesses . . . the gallery [in the] House [Parliament] would be deserted, and I used to come back and go over there on Sundays to work and Jane would send over food on a paper plate. I had a microwave in the office. Bolger, Birch and I and a couple of others used to work Sunday nights. Tipene would usually turn up on a Sunday evening and turn up for a long session and he smoked a . . . pipe and

I smoked cigarettes endlessly and we had a huge whiteboard, there was no other technology, and we would pace up and down and talk, generating smoke. We eventually worked forward . . . We had many discussions over quite a period of time.

What came through quite quickly was his view that last thing Māori really wanted to do was go back to court. Who knows what these . . . judges would decide! And I said we also did not want to go to court. So we actually found some common ground. We need to be very grown up about things because no matter what was going to happen there will be people opposing it from both sides, and we couldn't let the industry know because they would torpedo us and I didn't want the bureaucracy to know either, so he and I kept it quiet. So this was all about relationships. We trusted one another. Then the Ngai Tahu Fisheries report came out from the Tribunal and that had one simple recommendation, to negotiate, you have to negotiate. So we reached a point where we took the proposal to Bolger, it didn't really start until we took it to Bolger and we did that on a Sunday night. Tipene and I had worked out the concept as to how we might go about it and if one leaps right to the very end it turned out just how we wanted it.[2]

The government had set up a special task force to report on fisheries legislation and it released its report in mid-April. In the report, reference was made to exclusive Māori fishing areas (mahinga kaimoana) and also to the fact that the government needed to reserve 10 percent of quota for new species brought into the QMS in order to meet its obligations to Māori under the interim agreement. A nervous Sir Graham Latimer spoke against exclusive areas, saying that Māori could achieve control of their special fishing areas through the existing taiapure reserves provided for under current legislation (Maori Fisheries Act). These were not exclusive but still gave local control. Sir Graham felt that exclusive areas could create a backlash against progress already being made on Māori fishing rights.[3] Eastern Māori MP Peter Tapsell went further, claiming that setting aside areas of coast or sea for the exclusive use of local tribes was a recipe for disaster. It would cause dissension between Māori and Pākehā. Matiu Rata also criticised the report of the task force, saying that new fishery laws must not confine Māori fishing rights to reefs, rocks and picking pipi with a kit. He alleged that the task force's recommendations were diversions to avoid addressing the major issue of Māori fisheries. 'We are saying the Crown has a duty to restore to us not less than 50 percent of all the fisheries in all the categories,' he said. He noted that Māori were entitled to full ownership of the fish, but were prepared to gift 50 percent to the Crown.[4]

The Sealord opportunity

In early 1992, Carter Holt Harvey (CHH), owner of New Zealand's largest fishing company, Sealord Products Ltd, was under pressure. The company needed cash. In November 1991, CHH became an 'overseas company' when Brierley Investments sold its 32 percent shareholding in CHH to a joint venture which it owned 50/50 with the US International Paper Company. This put Sealord in breach of the 1983 Fisheries Act, which allowed only 25 percent overseas ownership of a fishing company without an explicit exemption from the Director-General of Fisheries. The then Director-General, Russ Ballard, eventually gave an exemption allowing 40 percent of Sealord to be overseas owned. The fishing industry complained bitterly.[5] However, what was not made public at the time was that the Ministry of Agriculture and Fisheries believed that Sealord had actually forfeited its fishing quota to the Crown as a result of operating for some time without this exemption. Sealord and CHH strenuously denied this, but a letter to CHH from Ballard on 2 July 1992, conveying the news of the exemption to the company, makes it clear that the quota had in fact been forfeited. It states: 'As you will see I have decided to restore to Sealord its quota holdings and permit it to continue to hold quota subject to conditions'.[6]

Fishing industry leaders were furious that government had granted special permission for the sale of up to 40 percent of Sealord to foreigners. The approval contradicted government assurances given in 1991 by Doug Kidd, who had said that the Fisheries Act did not allow foreigners to acquire quota. The fishing leaders were worried that allowing such a deal would see foreign fishing boats and crews coming back into New Zealand waters, as had been the situation ten years previously, when nearly all of our deep-sea fishing was done by foreign crews.

In early May 1992, when CHH announced its intention of quitting its shares in Sealord, the government had already agreed by letter in March to allow them to sell the shares, on condition that a single foreign investor would not be allowed to own more than 24.9 percent, nor to have more than one-quarter of the company's board members.[7] On hearing this news, Matiu Rata said he was seeking legal advice on how to challenge the government's decision to allow up to 40 percent of Sealord to be sold to foreigners. He was concerned that the government had still to deliver further quota to the Māori Fisheries Commission to satisfy the interim fisheries agreement.[8] In turn, Doug Kidd was keen to resolve the issue and get the suspended QMS working again so that all fish species could be brought into it. Māori believed that 50 percent of the fish quota would be a fair settlement, and this was the position they would take in talks with the Crown.

Behind the scenes, Māori interests were actively seeking ways of acquiring Sealord themselves. Sir Graham Latimer had tried to gather interests and finance to make a bid for the company; Tipene O'Regan had done the same, hoping to form a joint venture company with Royal Greenland (an Inuit owned fishing company from Denmark).[9]

Early moves toward Sealord

In March 1992 Her Majesty Queen Beatrix of the Netherlands visited New Zealand to celebrate the 350th anniversary of Abel Tasman's voyage to this country. Tasman was the first European to encounter Māori, and of course it was he who had renamed the country after the Dutch province of Zeeland. Tipene O'Regan, Sir Graham Latimer and Doug Kidd attended a reception for the Queen hosted by the government. During a break in proceedings the smokers gathered outside. Tipene O'Regan was chatting to Lady Emily Latimer, Sir Graham's wife, and Doug Kidd was there too. As Tipene tells it, he suggested to Doug Kidd, 'Do you want to resolve Māori fisheries?' Kidd was very keen, and so Tipene suggested to Sir Graham that in return for the Crown buying Sealord and handing it to Māori, they would agree to the settlement of Māori commercial fisheries claims. Sir Graham agreed and Doug Kidd was told.[10]

It was now up to Kidd to garner support for the proposition among his National Party colleagues, including Prime Minister Jim Bolger, who had rather conservative views when it came to Treaty of Waitangi matters. It appears Matiu Rata was not aware of these early moves being made by his fellow Māori Fisheries negotiators to stitch together a deal with government and acquire Sealord in return for settlement of Māori commercial fisheries claims. By July Dr Russ Ballard, head of the Ministry of Agriculture and Fisheries, was hinting that Sealord might be willing to sell some of its quota to the Crown to satisfy obligations to Māori.[11]

At the annual meeting of the Māori Fisheries Commission (Hui-ā-Tau) in late July, attended by 400 Māori representatives, there was robust debate between Tipene O'Regan and the other negotiators over Commission plans to allocate fishing quota. Matiu Rata recommended that none of the quota held by the Commission should be given to tribes until a final settlement was reached with the Crown. He and Bob Mahuta recommended that tribes lease quota in the meantime. They feared that acceptance of the interim 10 percent arrangement and subsequent distribution of that to tribes 'would "blow away" the negotiators' ability to succeed in court against the Government'.[12] Tipene O'Regan disagreed with this view, but acknowledged that the majority of the Fisheries negotiators opposed him.[13] A resolution from the meeting demanded that fisheries assets should be allocated as soon as possible. It seems from the discussions at the meeting that the Māori fisheries negotiators were keeping the prospect of acquiring Sealord and its 25 percent of commercial fish quota quite quiet.

Release of the Waitangi Tribunal Ngai Tahu Fisheries Report

The Waitangi Tribunal sailed to the centre of the fisheries stage again when it released its Ngai Tahu Fisheries Report on 6 August 1992. As with the Tribunal's release of the earlier Muriwhenua fishing report at a critical time in the debate over Māori fishing rights, the Ngāi Tahu report had a major impact. The timing was perfect. Doug Kidd came under immediate pressure. Speaking at the National party conference, he defended the Māori fisheries negotiators and the work of the Māori Fisheries Commission, saying that they had good business and economic growth acumen. 'Their stewardship of the quota and cash entrusted to them has multiplied substantially in a way that very few of our well known businesses have been able to achieve'. He also rejected recent criticism in Parliament of the Māori fisheries settlement by Eastern Māori MP Peter Tapsell, who had claimed Māori would squander income from fishing quota on 'huis and feasts',[14] basing this on what he said were the results of money going to rūnanga and the like in the past.[15]

Prime Minister Jim Bolger jumped in to downplay the Waitangi Tribunal recommendations on the Ngāi Tahu Fisheries claim by saying the government would not implement fully the recommendation of exclusive rights to most South Island fisheries. However, Bob Stannard of the Fishing Industry Board said that the Tribunal's report could destabilise fishing. He gave the example of the recommendation to return to Ngāi Tahu fishing rights in Lake Waihora, where there were existing licence holders, saying that this could mean compulsory acquisition, to which the industry was totally opposed. Another fishing industry leader, Peter Talley, described the Tribunal as a 'kangaroo court' and called for it to be disbanded.[16] The *Dominion* newspaper was predicting a 'Government caucus backlash' following the release of the Ngāi Tahu report.[17] Michael Belgrave, research manager at the Tribunal, wrote an article for the *Dominion* refuting various reports of the Tribunal's Ngāi Tahu fisheries recommendations. He showed that newspapers and television had exaggerated and twisted the recommendations, thereby causing unnecessary concern for ordinary people throughout New Zealand. He claimed that 'news media hysteria contrasted markedly with the measured responses of the tribe's leadership and the ministers responsible for Treaty of Waitangi issues', and pointed out that 'the tribe was not seeking all or most of the fishery'.[18]

With pressure at the approaching end of the 'interim settlement', the clamour from Māori for a just settlement, and hysteria from the fishing industry and the public over the Ngāi Tahu recommendations, senior government ministers were desperate to settle. In fact, they had already developed a basic agreement with Māori fisheries negotiators to purchase Sealord. This agreement had been thrashed out over about eight weeks 'behind closed doors' between the four Māori fisheries negotiators (Matiu Rata, representing the Muriwhenua tribes, Denese Henare, standing in for Bob Mahuta and Tainui, Tipene O'Regan for Ngāi Tahu, and Sir Graham Latimer representing the Māori Council) and the Crown, represented by Doug Kidd, Minister of Fisheries and Doug Graham, Minister of Treaty Negotiations, closely watched over by Prime Minister Jim Bolger and a number of government officials.

Who were the Māori Fisheries negotiators?

Matiu Rata

Matiu Rata was born at Te Hāpua in the far north in 1934. As a young man he was a leader of the Rātana youth movement, and he later became a seaman and trade union official. In 1963 he was elected as Labour MP for the Northern Māori electorate. Matiu was a stern and committed critic of National's Māori policies during the 1960s. He was Minister of Lands and Minister of Māori Affairs in the Kirk Labour government (1972–1975), and played a key role in drafting the Māori Affairs Amendment Act 1974. This major change in Māori land policy undid National's 1967 Amendment Act, which had been extremely unpopular among Māori because it provided for compulsory conversion of Māori freehold land with four or fewer owners into general land. It also increased the powers of the Māori Trustee to compulsorily acquire and sell so-called uneconomic interests in Māori land.[19] Matiu focussed on Māori retaining their land and using it for their own benefit. As Minister of Lands, he returned more Crown land to Māori control in a single term than any minister previously, but deliberately sought no publicity for fear of a European backlash. He said later, however, that 'No New Zealand citizen should fear the advent of justice for Maoris'.[20]

Matiu Rata earned a lasting place in New Zealand history as the man who was, more than any other, responsible for making the country address historic injustices to the Māori people. What is more, he did it with such charm and good humour that even the Pākehā (non-Māori) population respected and admired him. As noted earlier, he was also instrumental in establishing the Waitangi Tribunal in 1975. While he was Minister, the government also established Māori as an official language of New Zealand. Former Labour Prime Minister Sir Geoffrey Palmer said that Matiu was the catalyst behind the modern Māori renaissance. His close friend, Sir Graham Latimer, credited him with single-handedly being responsible for '70 percent of Maori achievements' in recent years that had, Latimer said, 'made an unequalled contribution to Maoridom'.[21]

The Labour government was defeated in 1975 and by 1979 Matiu had become dissatisfied with Labour's Māori policies. After being dropped as chairman of the Labour caucus committee on Māori Affairs, he resigned from the party in November 1979. He then formed his own party, Mana Motuhake (Māori Self-determination). He tested the goals of the party – Māori self-determination within a bicultural society – in a by-election in June 1980, but was defeated by the Labour Party candidate, Dr Bruce Gregory, and never succeeded in subsequent efforts to return to Parliament.[22]

Matiu Rata led the Far North (Muriwhenua) tribes in presenting their Treaty of Waitangi claims to the Waitangi Tribunal.[23] In 1988, when he was nominated as a Māori fisheries negotiator, he brought to the table a long and intense experience of achievement

NEXT PAGE

Hon. Matiu Rata at Manukorihi (Ōwae marae), Waitara, Easter 1983. He was attending a hui discussing pollution of the sea and shellfish beds from the government's 'Think Big' Synfuels project at Motunui.

Gil Hanly

against the odds for Māori rights. His leadership qualities stood out strongly, and his confidence made him want leadership. He was obviously single-minded and hard-nosed. He was a great organiser and relied on competent like-minded supporters for advice and encouragement. He was not considered extreme by the establishment and he did not quit. For example, in 1995, following the Sealord Deal, he was told to quit by his own people, but he ignored these calls.[24]

Bob Mahuta

In a tribute to Sir Robert (Bob) Mahuta, *Tangaroa*, the official magazine of TOK noted that he loved the song 'I did it my way' (made famous by Frank Sinatra) and that he lived by those words.[25]

Robert Ormsby was born in Te Kuiti in April 1939, and adopted at four weeks old by the Māori King, Koroki. He grew up with the royal family on the Waahi Marae, where his first five years were spent living in a raupō hut with dirt floors. Te Puea (known as Princess Te Puea, grand-daughter of the second Māori King, Tawhiao Te Wherowhero) had Bob sent to Mt Albert Grammar School in Auckland, which he left having failed the national examination, School Certificate.

He then worked in a series of labouring jobs – first coal mining, then the army, the wharves and the freezing works. Perhaps Te Puea had Bob in mind for succession to the King. However, he told Paul Diamond, 'I personally didn't aspire to become Koroki's successor. It had been touted when I was young, still in my twenties, because I was going to varsity at Auckland, but it was all bloody press talk'.[26] At the age of 24 he changed his name to Robert Te Kotahi Mahuta.

Certain leadership personality traits were already evident: 'Even at that young age, I showed pig headedness – that was what I wanted to do and I was going to do it, and nobody – including Te Puea – was going to tell me otherwise. I'd get there, but in my own way, not by anybody else's'.[27]

My wife's experience with education had a major influence on me – that, if she could do it, us other Maoris could do it. Even though I had no qualifications, I said, if they can do it, I should be able to do it. Up until I was fifteen, I was getting education thrust down my throat all the time; I was getting sick of it. I just wanted to get out of it, so I finished school and went straight into the mines. Anyone who has been in the mines knows it's bloody hard work. It's grinding day-in-day-out work that seems never ending.[28]

. . . Being in the army taught me a lot about life. We were taught to do menial meaningless tasks at the drop of a hat; that there's no short-cuts to success or victory; that whatever you want to do is a challenge and that challenge must involve hard work, and you mustn't be afraid of it. I'd already experienced five years of hard work in the mines; I didn't want more of that, where there was no future through it. So, I thought the army was another way of working hard.[29]

Raupatu [the wrongful confiscation of Tainui land in the 1860s after the government had sent an invasion force into the Waikato] too was a cause, in a sense, worth dying for, and if I was going to putt out, then I wanted to do it for something that was worth dying for. I was afraid of nothing after that.[30]

After marrying, Bob studied at night school and at the University of Auckland, where he attained a master's degree in anthropology. He then

TOP

Bob Mahuta (left) with Dame Te Atairangikaahu, Prime Minister Jim Bolger and Doug Graham, following the signing of the Waikato Deed of Settlement at Tūrangawaewae, 1995. Bob Mahuta (later Sir Robert) is one of the heroes of the fisheries story. His and Tainui's support for Muriwhenua and Ngāi Tahu in challenging the Quota Management System were crucial to the final outcome.

Mahuta Family

LEFT

Prime Minister Jim Bolger and Ngāi Tahu leader Sir Tipene O'Regan hongi after signing the Ngāi Tahu deed of settlement at Takahanga Marae in Kaikoura, 21 November 1997. Treaty Settlements Minister Doug Graham signs at left. Sir Tipene had by this time established himself as a visionary and pragmatic leader as chair both of TOK and of Ngāi Tahu.

Fairfax NZ: Peter Meecham, Signing_2_211197_PM_A_90679
The Christchurch Press

lectured there in Māori studies. In 1972, aged 33, he became director of Māori studies and research at Waikato University. He then studied at Oxford University. While he was there, in 1977 'the English banned him from playing rugby union for life after the post-graduate student was discovered breaking the rules by playing rugby union on Saturdays and league on Sundays. He laughed off the ban as ludicrous'.[31]

Bob was called home to deal with a major project that seemed to compromise Tainui rights – the Huntly Power station. The station was designed to burn coal and discharge heated water into the Waikato river, held to be sacred to Tainui.

> I can't pick when the spirit of activism began to enter us, to take a more proactive view towards the 1946 settlement. We younger leaders had confrontations with Pei Te Hurinui and the old Tainui Maori Trust Board because they reminded us of what Te Puea had said, and what she'd agreed with Fraser [New Zealand Prime Minister]. To them, what we were trying to do was counter to Te Puea's agreement. I said no, no, no, that was during her time, this is our time and we've got to do it another way, there must be another way to do it. We could sense that there was a sea change in attitudes toward these grievances.[32]

In 1988, the Labour Government wished to sell the government owned Coal Corporation. Tainui challenged the proposal in court and, by doing so, were able to raise the issue of the earlier raupatu (confiscations) and the 1946 settlement (so-called). The Court of Appeal found in favour of Tainui.

> It stated that a considerable portion of the lands confiscated in the 1860s had been taken in breach of the Treaty of Waitangi. It also held that coal mining rights were a valid part of Tainui's claim and that these were eligible for inclusion in any subsequent settlement. The President of the Court, Sir Robin Cooke, said the payment agreed as part of the 1946 settlement was 'trivial', and that 'some form of more real and constructive compensation is obviously called for if the Treaty is to be honoured'. He also said that Tainui could return to court if the Government failed to negotiate. The decision laid the base for the negotiations between Tainui and the Crown, which would culminate in the 1995 settlement with the National Government.[33]

Bob Mahuta was instrumental in and probably critical to finalising this settlement.

At that time Tainui became part of the Māori fisheries case, adding their might to the opposition to government plans implementing the Quota Management System (QMS). Bob was appointed one of the Māori Fisheries negotiators and his role was taken up by Denese Henare.

Tipene O'Regan

Tipene O'Regan was interviewed by Paul Diamond for a book on leadership, *A Fire in Your Belly,* in 2002.[34] Born in Wellington in 1939, he told Paul that his father (Roland O'Regan) was a strong influence, 'because he freed my thinking. He gave me confidence and he gifted me with a love of wrestling with ideas and of trying to find the solutions to big problems. The thing that makes me different is that I also enjoy small problems – like how to fix and make things. I am the beneficiary of his personality and his freedom of intellect'.[35]

The O'Regan family home was ransacked by policemen in 1951 when Roland was Chairman of the Roseneath branch of the Labour Party. Tipene recalled the state of terror he found his mother in when he arrived home from school. 'I learnt early that, on the whole, the state is the enemy of the citizen.'[36] His father and his uncle Con were deeply involved in the issues of what was essentially Irish autonomy. 'I transferred a lot of that into my perceptions of what I found facing me when I got called to the frontline within Ngai Tahu. Until that time, my Ngai Tahu world was one of kinship, the coast, the sea, people I knew and loved, and the issues confronting Ngai Tahu, as a people, didn't hit me until then, when I was in my twenties'.[37]

Tipene graduated from Victoria University and later worked as a lecturer at Wellington Teachers College until 1983. In 1976 he became a member of the Ngai Tahu Maori Trust Board and remained there until 1996. He became a Māori fisheries negotiator in 1988.

He referred to the signing of the Ngai Tahu Deed of Settlement with the Crown in 1996, in recognition of the tribe's numerous grievances, as follows: 'I think the day I delivered Ngai Tahu the opportunity to have a future, not so much with the passages of legislation, but the Deed of Settlement at Takahanga Marae in Kaikoura – standing alongside my beloved friends, Bill Solomon, Charlie Croft, my family and the other tribes that came (the old man from Tainui, and his delegation, and the people from the other iwi who were there). The memory of that day stands on its own, despite the fact that my people refused, or failed to grasp, the real potential in that settlement'.[38]

'If you were to ask me about the nature of leadership, in terms of what I've learnt, I'd say you've got to have a fire in your belly for an outcome. On the whole, I don't think people trust leaders who are only interested in personal power. Leaders have got to have a sense that what they want is what the people want.'[39] Indeed, a Māori Land Court Judge, Heta Hingston, commented that Tipene had an 'overbearing manner' when dealing with dissenting shareholders of the Mawhera Incorporation.[40] He is a big man and does have a strong, domineering personality, which, combined with his strong voice, quick wit and grasp of facts, is a challenge across any negotiating table. These are the attributes he brought to the Sealord negotiations in 1992.

Graham Latimer

Graham Latimer was born in 1926 in the Far North (Tai Tokerau). He attended Kaitaia District High School, but was required to help on the family farm and left after two weeks. He said in his biography, written by Noel Harrison, that the Great Depression had a big influence on him.[41] He worked on farms until enlisting in the Army in 1943, and went to Japan in 1946–47 in Jayforce as part of the army of occupation after World War II. Graham then worked on the railways, married and began farming around the Kaipara district.[42]

Graham was elected to represent the Tai Tokerau district on the New Zealand Māori Council in 1964 and became national president of the Council from 1973. In his capacity as president, he undertook certain legislative responsibilities for the welfare of Māori in general that were specified in the Act of Parliament governing Maori

Māori Council chairman Sir Graham Latimer at the council's annual meeting at Te Puni Kōkiri, Wellington, 25 June 2003. Sir Graham's life has been devoted to securing a rights-based future for Māori through the Treaty of Waitangi.

Fairfax NZ: Craig Simcox, Latimer_1_2506_25751 Dominion Post

Councils, the Maori Community Development Act 1962. Section 18 states that, under 'General functions of the New Zealand Maori Council', it is charged with assisting Māori . . . 'to conserve, improve, advance and maintain their physical, economic, industrial, educational, social, moral, and spiritual well-being . . . In the exercise of its functions the Council may make such representations to the Minister or other person or authority as seem to it advantageous to the Maori race'. Many Māori believed that this legislation was patronising and outmoded, but Graham Latimer was to lead the Council to success in several high profile legal challenges against government actions. The Māori Council had been for many years officially recognised by government as representing all Māori, despite what Māori thought about that.

In 1977 Graham was appointed to the Waitangi Tribunal. Until then he had not been well-known to the general public, although he was extremely well-known within Māori communities. In Tai Tokerau he was acknowledged to be a strong advocate for the north, with unequalled political connections in Wellington. He was knighted in 1980 and was elected Māori vice-president of the National Party in 1981, remaining in that position until 1992. 'Within the higher ranks of the National Party he was a familiar figure, seen as an astute operator, not likely to rock the boat, and useful as a spokesperson in a party with a long record of distancing itself from Māori issues.'[43]

Latimer's position as president of the Māori Council had been disputed for many years. While he was respected for helping to keep the Council alive, opponents criticised him for failing to win greater self-determination for Māori.[44] Noel Harrison said of him:

> He did not have a charismatic persona, instantly recognisable on television. He was not a celebrity. Few of the many organisations he worked with generated headlines. He did not demonstrate yearly at Waitangi, but sat with dignitaries while some of his friends, relatives and enemies angrily abused the establishment for oppressing Māori, and him, for sitting with them.[45]

Wira Gardiner, former chief executive of Te Puni Kōkiri, said during the early 1990s that Graham Latimer was one of the most important Māori leaders: 'He's taken advantage of opportunities and he's succeeded where others haven't acted.' He also said that Graham was not afraid of failure, had the political experience to know when to try something, and had 'an impeccable capacity for opportunism, is a supreme optimist, and has been the driving force behind numerous developments'.[46] Once Graham Latimer was appointed as a fisheries negotiator representing the Māori Council, his close connections with the National Party government in 1992 made him extremely valuable to the negotiating panel. He demonstrated strong leadership as a director of Aotearoa Fisheries Ltd, and he demonstrated canny vision when he told Adam Gifford, 'I asked all the directors [when he was chairman of AFL] what their ambition was, and why did they come and sit on the board of AFL. And I let everyone have their say and I said to them, "Oh, I have an ambition that the Māori will be the biggest fishers in this country, or the biggest company" . . . and they said "What, in Auckland?" And I said, "No, in New Zealand"'.[47]

The Crown negotiators

Doug Kidd

Doug Kidd was born in Levin and grew up on the family dairy farm at Kuku, just south of Levin. 'Living in a rural community where my playmates were Māori kids helped me build some understandings which would stand me in good stead many years later.'[48] Pākehā were 'the minority culture in our school and in the community'.[49] From 1960 to 1964, he served in the New Zealand Army territorial force. He later obtained a law degree from Victoria University of Wellington.[50] Working as a young lawyer in Blenheim, he handled many cases for Māori land owners. He describes Māori as part of the fabric of his life.[51] He also had business interests in aquaculture, forestry, and wine making. Doug Kidd was a driven businessman, and in Marlborough he fought hard for the development of the region. He was a strong advocate of commercial forestry, and was a member of the Marlborough Catchment Board that also strongly promoted forestry. He was elected to Parliament as MP for Marlborough in 1978, and later became Minister of Fisheries in the Bolger government. Doug Kidd was astute and forceful and not frightened to try something new. His openness and pragmatism were important as was his directness when dealing with people.

Doug Kidd was a member of Parliament's Primary Production Select Committee, when it reviewed the 1908 fisheries legislation that was to become the Fisheries Act 1983. In his words:

Discussion centred that day on the question of whether to remove the old provision that nothing in the Act would affect Māori fishing rights. My recollection is that I was the only one, or virtually so, in the room in favour of retaining the provision. It is one of those strange ironies in life that it later fell to me to negotiate to buy out that provision during the making of the Sealords Deal.[52]

In October 1991 Prime Minister Bolger asked Doug Kidd to take on the role of Minister of Māori Affairs after Winston Peters had been sacked:

My background as a dairy farmer's son, part-time mussel farmer, chair of local wine and forestry companies, President of the Marlborough District Law Society, foundation President of the Marlborough Forest Owners Association, elected member of the Marlborough Catchment Board . . . gave me a broad appreciation of land and resource issues which I was able to apply when I became Minister of Māori Affairs.[53]

On becoming Minister of Māori Affairs, Doug Kidd was faced with three major issues: modernising Māori land legislation that had last been reformed in the 1950s and was desperately in need of revision to reflect contemporary Māori society; resolving the greatly inequitable issue of Māori reserved lands – particularly in Taranaki – where such lands were Crown owned and leased to (mainly) non-Māori farmers paying peppercorn rentals; and finding a resolution to the Māori Fisheries dispute.

Doug Kidd recalled that when he became Minister, he was 'amazed that so little had been

done in the fifteen years since Dame Whina Cooper had led the great hikoi' [Māori land march].[54] 'I proposed that we get a group of radical hot-headed young Maori lawyers – the likes of Annette Sykes, Joe Williams and others – to draft a Bill. The challenge I put to them was: 'You write it; I'll enact it and you live with it'. Within two years, with Doug Kidd 'driving' government officials, the Ture Whenua Maori legislation was produced, completely revising the old Māori land legislation and becoming law in 1993. One of the first steps he took demonstrated his pragmatism and desire to get things done.[55]

There are detailed accounts of Doug Kidd's achievements as Minister of Māori Affairs in *Tui, Tuia – Reflections on Māori Development 1984–2004*.[56] For the purposes of the Māori fisheries story, his life experiences, his personality and his friendship with Tipene O'Regan were all of crucial importance. Later in 1996 he was appointed Speaker of Parliament, and in 2004 he was appointed to the Waitangi Tribunal.

Doug Graham

Described by Ruth Laugesen in a *Dominion* newspaper article as, 'the distinguished-looking plummy-toned National politician with the air of one born to rule',[57] Douglas Graham grew up in Auckland. He graduated with a law degree from the University of Auckland, establishing his own law practice in 1968. From 1973 to 1983 he lectured in legal ethics at the University of Auckland. He was elected to Parliament in 1984 as MP for the Auckland electorate of Remuera. When the National Party won the 1990 election, Doug Graham was appointed to Cabinet, becoming Minister of Justice, Minister for Disarmament and Arms Control, and Minister of Cultural Affairs.

In 1991, he became Minister in Charge of Treaty Negotiations, perhaps his most prominent role. He was widely praised by both Pākehā and Māori for his work on numerous Treaty settlements, although opponents of the process have voiced criticisms of his policies. Later, he also became Attorney General and Minister for Courts. In the 1996 elections, he became a list MP. He retired from politics at the 1999 election.[58]

Doug Graham's contribution to the Māori fisheries negotiations was as a support for Doug Kidd. He knew the law, and although he did not know a great deal about Māori claims when he became Minister for Treaty Negotiations, he learnt fast and was successful. Later settlements he achieved (for example, the 1995 Tainui settlement and 1996 Ngāi Tahu settlement) were significant and set benchmarks for all major settlements that followed. These demonstrated that he too was pragmatic, driven to achieve the goal.

In the Māori fisheries story, Doug Graham played a supportive role. The Crown team of government officials and Doug Kidd, with the backing of Doug Graham and Jim Bolger, made a cohesive, pragmatic and somewhat empathetic team. The government needed a settlement and Māori wanted one.

OPPOSITE PAGE

Doug Kidd at Parliament, c. 1994. He has been crucial to the Māori fisheries story, as one of the most active Fisheries and Māori Affairs Ministers.

Fairfax NZ

So the scene was set late in 1992. On the surface, Matiu Rata, Sir Graham Latimer, Tipene O'Regan and Bob Mahuta were pitted against the pragmatic and entrepenuerial Minister of Fisheries and of Māori Affairs, Doug Kidd, and the Minister for Treaty Negotiations, Doug Graham, with Prime Minister Jim Bolger overseeing the unfolding Māori fisheries drama.

A great deal was at stake. From the government's perspective, an equitable and amicable settlement of the Māori fisheries dispute, which had festered for many years, would restore honour to the Crown, signal that they were serious about settling historical grievances, and restore confidence in the fishing industry as a whole during a period of far-reaching change. Digging deeper, there was more than a little goodwill between the negotiators, and nobody relished the prospect of going back to the Courts for a solution. There was a looming resolution to the issue that was positive for all parties, but the negotiators had to be big enough and bold enough to seize that settlement.

OPPOSITE TOP

Prime Minister Bill Rowling greeting Whina Cooper, leader of the Land March, Parliament, October 1975. By 1990, Doug Kidd, newly elected Minister of Māori Affairs, was shocked that in the fifteen years since the Land March, nothing much had been done to give Māori land more legal protection. He was determined to do something about that.

Newspix.co.nz: 1080162 NZ Herald.

OPPOSITE BOTTOM

Ngāti Ruanui land claim settlement deal being signed at Parliament in September 1999. From left: Treaty Negotiations Minister Doug Graham, Georgina Te Heuheu, Pat Heremaia, Maimau Maruera and Steve Heremaia. Doug Graham strove to settle as many Māori Treaty grievances as he could during his time as Minister.

Fairfax NZ: Don Roy, 30-dpt-Sept1999-250 Dominion Post *archive 040112*

ENDNOTES

1. *Dominion*, 27 March 1992. p. 3.
2. Doug Kidd, interview with Brian Bargh, November 2014.
3. *NZ Herald,* 21 April 1992.
4. *NZ Herald,* 28 April 1992.
5. Bill Rosenberg, 'Fisheries giant could come home.' *The Press*, 5 May 2000.
6. Ibid.
7. *NZ Herald*, 4 April 1992.
8. *NZ Herald*, 15 May 1992.
9. Tipene O'Regan, pers. comm., 2014.
10. Ibid.
11. *Dominion*, 8 July 1992.

12. *Dominion,* 27 July 1992.

13. Ibid.

14. *Dominion,* 10 August 1992. p. 2.

15. *Dominion*, 25 July 1992.

16. *Dominion,* 12 August 1992. p. 1.

17. *Dominion,* 13 August 1992. p. 2.

18. *Dominion,* 18 August 1992. p. 6.

19. http://www.nzhistory.net.nz/people/matiu-rata

20. http://www.independent.co.uk/news/people/obituary-matiu-rata-1253022.html

21. http://www.independent.co.uk/news/people/obituary-matiu-rata-1253022.html

22. http://www.nzhistory.net.nz/people/matiu-rata

23. Ibid.

24. Margaret Mutu, *The State of Māori Rights*. Huia Publishers, 2011. p. 27.

25. *Tangaroa,* No. 59, March 2001. p. 3.

26. Paul Diamond, *A Fire in Your Belly: Māori Leaders Speak*. Huia Publishers, 2003. p. 120.

27. Ibid.

28. Diamond, 2003. pp. 121–22.

29. Ibid, p. 122.

30. Ibid, p. 142.

31. http://www.nzherald.co.nz/nz/news/article.cfm?c_id=1&objectid=170805

32. Diamond, 2003. pp. 123–24.

33. Ibid, p. 134.

34. Ibid, pp. 9–50.

35. Ibid, p. 16.

36. Ibid, p. 19.

37. Ibid, p. 27.

38. Ibid, p. 43.

39. Ibid, p. 41.

40. *Dominion,* 15 August 1992.

41. Noel Harrison, *Graham Latimer – A Biography*. Huia Publishers, 2002. p. 19.

42. http://en.wikipedia.org/wiki/Graham_Latimer

43. Harrison, 2002. p. 225.

44. Ibid, p. 225.

45. Ibid, p. 184–85.

46. Ibid, p. 189.

47. Adam Gifford, interview with Sir Graham Latimer, 1995.

48. Te Puni Kōkiri. *Tui, Tuia – Reflections on Maori Development 1984–2004*. Te Puni Kōkiri, 2004. p. 42.

49. Maria Bargh (ed.), *Māori and Parliament; Diverse Strategies and Compromises*. Huia Publishers, 2010. p. 189.

50. Doug Kidd, pers. comm., 2014.

51. *Dominion*, 6 October 1992.

52. Te Puni Kōkiri, 2004. p. 44.

53. Ibid.

54. Ibid, p. 45.

55. Bargh, 2010. p. 191.

56. Te Puni Kōkiri, 2004, pp. 44–7; Bargh, 2010, pp. 189–97.

57. *Dominion*, 6 October 1992.

58. http://en.wikipedia.org/wiki/Doug_Graham

HAULING IN THE CATCH:
The negotiations' aftermath

Sir Tipene O'Regan, Sir Robert (Bob) Mahuta and Sir Graham Latimer were interviewed by journalist Adam Gifford about the Sealord deal towards the end of 1995. The author interviewed Sir Doug Kidd in November 2014 and Sir Tipene O'Regan in April 2015. Each man had his own view of the negotiations. It is clear that both the Crown and the Māori negotiators were under time pressure. Everyone knew that Carter Holt Harvey was keen to conclude a sale of their shares as soon as possible, and there were already indications in July 1992 that another consortium was preparing to bid. There was also pressure from the fishing industry and Māori for a settlement that would bring certainty to the industry. Much was at stake, and so the negotiators devoted only eight weeks to reaching agreement. Doug Kidd recalled the brief and hectic negotiation period in his 2014 interview:[1]

Carter Holt Harvey decided they would sell out of Sealord and it was just like something had come down from heaven, amazing timing. The first thing that happened after I talked to Bolger was he said you've got to involve the other claimants so that's Sir Graham on the Māori Council, Robert Mahuta and Matiu on behalf of Muriwhenua, and they had of course had their fisheries report that nothing came of, and then of course Tipene was Ngāi Tahu. We had just got into, late the previous year [1991], the idea of doing something serious with Treaty settlements. Doug Graham was appointed Minister of Treaty Settlements. We had formed a Cabinet committee that I was on, not surprisingly, because I was the Minister of Māori Affairs. The PM can go to any meeting and often did. But that committee had nothing to do with the Sealord deal. And Doug Graham truthfully said to the boss, 'I don't know a kahawai from a kingfish, he'll [Doug Kidd] have to do it.'

So we rounded things up, set up negotiations and they were done in my office on the sixth floor – a very big office, very long table and they all got around it. Whole lot of people turned up and they all came with lawyers . . . and suddenly I felt now I was losing it. We couldn't move at this pace because we were nit-picking over every word and syllable and I said, 'Right, this is not going to work. We are going to have to get rid of the lawyers.' And they all moaned and I said, 'Well we are. This is going to have to be chiefs to chiefs and all the others out.' And they all went out with their tails between their legs and we got going again. Sir Graham, Robert Mahuta who was soon to fall sick, I think within hours he did, I think he got to the first session but . . . Denese Henare appeared thereafter, for Tainui. Matiu was there and Tipene.

Of course I had been sent down to 'assist' me, in other words to spy on me, two guys out of the PM's department, names of Shane Jones and Peter Douglas. Two bright and upcoming young fellows – fresh young fullas just back from Harvard, new wave, crème of the crop of young up-and-coming Māori, and they were enormously helpful. They worked as my mustererers . . . We couldn't have smoking in this room because it just would have killed us so we used to have smoke breaks and of course they were good, very good. They (the Māori negotiators) would get out and hold themselves together. They had trouble holding themselves together. When Denese came to us . . . I don't think she knew anything about it . . . she felt the burden of Tainui . . . I think it was a huge burden. But she was there at the end, some didn't make it. Somehow we had deputy negotiators who got into the room at times – Dick Dargaville and Whatarangi Winiata . . . a lovely gentleman, and a couple of others. But the truth was we had the four main ones and I think I've mentioned to you we had these peculiar coincidences with relationships. Bolger and Latimer were thick as thieves from the National Party. Then Bolger and Mahuta were the couple of Waikato characters – Waikato rugby and other stuff . . . who knows what goes on in and around Te Kuiti. Matt was the odd-ball and he was quite difficult to manage but the smoking breaks were very helpful for that . . . he talked up the Māori case and he wanted to be sure that Māori could get to 50 percent. He didn't expect to get it at the table but he wanted the way to be open to get to 50 percent and I think in the end he was content.

We got through all that, then there was this very serious problem; what were we going to tell Cabinet. I was totally exposed, totally exposed. But you know I don't recall any sense of fear about that.

Doug Kidd said that he relied on close support from:

. . . Phil Major, who was my chief of policy and very important in all of this. It was one of those cases of an official and a minister, me towards him, having total confidence in him. He had a good mind. Great guy, a good guy in thinking with whatever the problem was, he must have been trained in policy. At times, and it was probably later on, Terry Lynch served me well too, especially with eels and things.

In addition, there was also the required support from Doug Graham and, less frequently, Jim Bolger. But there was no-one from Te Puni Kōkiri, of which Wira Gardiner was Chief Executive. When he found out about the negotiations he was angry, and Doug Kidd comments:

I had no excuses. I never rejected him. I couldn't stop the . . . thing to say, 'Hang on, I want to get someone else on board, who would argue everything.' But I come back to, because I had those two portfolios, I had total power and equipment, standing, to do this. And those relationships formed the scrum and it was kicked around a little bit and we got back up, but by and large it went towards the line. I had another image of how it all happened because I have quite a soldiering background, but it seemed to me to conjure up an image of the Somme. Everybody in deep bloody trenches, going nowhere . . . anyone who put their head up would get shot. It was a case of Tipene and me meeting up in no-man's-land in the middle and working. The main risk is we [would] get killed by our own side. If they knew we were there, they would have. There was so much baggage you really had to get out of your trenches and into this dangerous position and do it and invent it. We made it up and I don't apologise. With propellant and good fortune seeming to smile on us, it's almost impossible to write anything instructional for those who follow; it was a product of its time, desperate need in fisheries terms.

We got to the point of when we had to take it to Cabinet. Well of course the papers go out on Friday and the National Cabinet are all over the landscape. These things were sent by official cars to post offices and homes and milking sheds and then everyone reads it and goes, 'What the . . . is this!' My phone starts ringing, 'What the hell have you been up to? Are you trying to wipe the National government?' Very direct sorts of folks but no nastiness, they knew this was big stuff and we had had big stuff; Bank of New Zealand fell over just before we were sworn in so we had dealt with big things. So we get to Cabinet and God bless Jim Bolger, sat there solid as a . . . rock he would not be moved. Pretty soon the Cabinet got the feel; it would be unfair to say Cabinet had some red-necks. Couldn't possibly say there were none of those. But they decided to give it a go and we were working on a high majority of Cabinet agreeing, not just one – Bolger didn't work like that. We worked on consensus. We probably spent as much time talking about how we were going to take this to Caucus because we had a huge caucus of course with the overwhelming majority. Birch was also right on side. Bolger and him were as thick as thieves. So we went to Caucus and it would be fair to say there was some room between the floor and their feet. Well, the faces went red, and caucus engaged vigorously in the subject. They all said they'd lose their seats.

ABOVE

Dr George Habib (left) talks to Māori Fisheries negotiator Matiu Rata during Sealord fishing deal briefings at the Beehive, 24 September 1992. Habib's reports on Māori fishing were extremely valuable to the Waitangi Tribunal for its findings in the Muriwhenua Fishing Report.

Newspix.co.nz: 1080160 Martin Hunter, NZ Herald

RIGHT

Bob Mahuta in reflective pose at Coalcorp, Huntly, 1989. Denese Henare attended many of the fisheries negotiation sessions on Bob Mahuta's behalf. However, her instructions came from him and he in turn kept Tainui fully informed.

Mahuta Family

But in the end Bolger's leadership prevailed. I was the skunk in the room. I don't think I have ever been forgiven by some of them, including some of those that are in power now.

Anyway, Caucus got it through, not by vote – I can't remember ever having to vote – just by consensus. Sometimes things would be sent back for further work. Well, everyone went away to prepare bomb shelters, I suppose, so then we had to let it out into the open. We got this heads-of-agreement sorted out with the negotiations so we called a meeting with Māoridom, and just think how simple and easy communications were not that long ago, anyway they came in bus loads. And people got up and spoke . . . There was some fierce talk and the more uninformed they were the fiercer they were. We had to carry the day. I was never so humbled as when my kaumātua from Ngāti Koata/D'Urville Island – Pene Ruruku, very quiet . . . he stood up and walked forward and when they invited people to sign he stepped forward. I thought 'how brave'.[2]

Bob Mahuta recalled that negotiations began with a meeting of the negotiators about eight weeks before the deal was agreed, to talk with a potential funder for a bid to buy Sealord–Bancorp:

About a month or so later, Graham [Latimer] rang up to say we should go and see Bolger to see whether the government was going to support this. So we went down and Graham and Steve [O'Regan] were there. And I said, 'Where's Matt?' And they said, 'The government doesn't want Matt anywhere round the place.' And I said, 'That's bloody silly. Why are they bringing their dislikes of Matt into this? We're just talking about a pan-Māori issue.' Anyway, we went in and sat down and Bolger offered whisky and that, and Steve and Graham drank and I wouldn't touch it. And then we broached this question about the Crown, the Crown providing the capital for us to buy Sealord. And we were skirting about how much we needed and then Steve said 'Oh well', Steve was getting a bit drunk by this time and said, 'Oh, we can do it with $50 million, eh.' And I thought 'Where the hell is he coming from?' We named numbers because we knew it was $170 [million].

And anyway we managed to quieten him down and then we ended up with the number, 170. And Bolger got Bill Birch [a senior member of Cabinet] in and he said, 'These fellas want $170 [million] to buy Sealord.' And Birch said, 'Oh yes, that looks all right. We'll look at it in more detail.'

Then after that it was hard negotiations at the Portland Towers [a Wellington hotel], up at Parliament, and to-ing and fro-ing. I've got a diary note there about it but this was the time that Steve was under a lot of stress because his father was very sick.[3]

Negotiations continued, and within a day or so, as Bob Mahuta recalls:

We were pretty well close to it then. And so we retreated to another side room just to get the details right, and I chaired that last session. And I said, 'Well Matt, we're getting closer to clinching this deal, it's a Treaty deal, and my view is that whilst it's not an ideal settlement, at least it gives us the opportunity to make the Treaty level and that in view of that I thought that perhaps we should have a bit of a karakia.' Matt said, 'You're the Morehu, the Minister, you should take the karakia.' OK, so Matt he agreed to that and so we had the karakia there and Steve of course was still holding out for some side deal and said well, he wasn't going to agree to it, eh, even after the karakia. He said he had to go for a walk. And I said, 'Well that's fine. You go for a walk. You can walk as far as you like. But when you come back we just want a yes.' So he went off. And so we waited about 15 minutes and he came back and said, 'Well, I've weighed it up. I don't like it but I'll say yes to it. But I want it on record that I don't like it.'[4]

Bob Mahuta believed that Tipene O'Regan's 'side deal' related to the allocation of pre-settlement assets (to be allocated on the basis of a tribe's coastline, which would have favoured Ngāi Tahu), and mahinga kai areas (customary fishing areas) to be set aside. Because these things were not agreed, they were omitted at that stage. Asked to comment by the author in 2015, Sir Tipene said that the issue worrying him was that the negotiators had strayed away from a rights-based approach:

> Well, it was the nature and the right . . . customary rights . . . The question was that the customary right in fisheries has always been of enormous importance. I know it is for Māori generally, but it is right at the cornerstone of Ngāi Tahu identity, it is the whole mahinga kai case. People often don't understand the enormous significance of the tuna fishery and those others, because when we lost all our lands we basically kept our people together on a basis, our whakapapa rights and everything all locked up in mahinga kai. So, it is always at the heart of Ngāi Tahu consideration. That is why we are so strong on environmental matters and things of that kind now.
>
> Now what was missing was the prospect of a Treaty fishery settlement in which all those values related to traditional customary non-commercial subsistence rights were basically just being brushed aside. We were going to end up no different from a Pākehā recreationalist. The lawyers and Brierley's were only interested in the commercial deal. The whole thing was absolutely being sorted out in terms of money, yet there was this bundle of values represented by rights in fish …[5]

Graham Latimer recalls:

> We just had heard by accident that Sealord was up for sale, probably Tipene and Doug Kidd could have talked about it, I don't know. But I talked about it with Rob Challinor and Wayne Boyd [both professional company directors]. After we put the idea we went down to see the Prime Minister and he just gave us a flat no, and as I was walking out the door, I said, 'Does that mean that the door's closed?' And his response was, 'It's never closed to you, Graham.' And I think Matt can verify that.

Sir Graham believed that Bolger would only listen to a commercial argument from Māori. He consulted with Bancorp (presumably on behalf of Aotearoa Fisheries Ltd, of which he was Chair): 'Matt stayed on his trail, trying to get it [Sealord] for nothing, and that was what his job was . . . and Tipene got tied up with Royal Greenland [a Danish Inuit company]'.

Shortly after that, Graham Latimer recalls, a meeting was held at 'Anzac Avenue' with Matt Rata, Bob Mahuta and himself (Tipene O'Regan was not there), and they agreed to try to purchase the company. He flew to Christchurch to let Tipene know what they were planning. Sir Graham continues:

> It was then that Matt and them made a decision that I had to fly to Christchurch to put our Sealord deal to Tipene. When I got there Tipene and Robin Hapi and Whai Dewes were there . . . Tipene said, yes, well, he'd agreed to our way providing he was Chairman [probably of Sealord, of which Tipene was the founding Chair from 1993 until 2002]. And of course you're in another man's territory. You had to concede. Now when we got everything going, because by that time I'd gently felt my way around, and the general consensus of opinion was that Māoris couldn't run a big business – that was [what] really put paid to the political distance, so Rob Challinor and Wayne Boyd tried all the big fishing industries, tried overseas to see whether they could come in and take a 50 percent position.[6]

The negotiators then had to find the funding to finance the purchase. Sir Graham and Bancorp tried numerous avenues to raise the funds including overseas fishing companies. Sir Graham was certain that the government was doubtful about Māori owning the company alone.[7] The Māori negotiators had to consider whether Māori would buy the company alone or in partnership. Graham Latimer continues:

> We tried Brierley's and I went down to see Bruce Hancock, and he can tell you, I was the first one ever to make approaches. And his first response was, 'What the hell would Brierley's want to be dealing with Māoris for?' I thought . . . Anyhow he came back. He went away and thought about it, and came back and thought about it, and we

decided that we'd try and get a JV [joint venture] through, and get the government to agree. So as you know, Bob and I and Tipene went up and the Prime Minister said, 'If you can get Brierley's in, you've got a deal.' And that's really how it came about. By that time, Matt had come into [it] . . . Matt's strength in the political world really showed up at that stage . . . Matt could spot the political side from a mile away and he'd say, 'No, you can't go that way. You have to go this way.' So there was a lot of . . . Matt's leadership was quite good, I thought, at that stage.[8]

Sir Graham Latimer also recollected the final negotiations:

They [Doug Kidd and Doug Graham and their advisors] had to negotiate the price. Remember that we had $150 million, and if it was going to be beyond that we never had a deal. We could come in under. We had it pitched there. So I went up and saw Matt with the Prime Minister, we promised $50 million a year for three years, and next morning we went in, they were going to put GST on it, and we had a hell of a row and we brought the Prime Minister down and I said 'Last night we said $50 million a year, net; no GST.' And he said, 'That's right, no GST?' Treasury nearly died. So really we got $168 million.[9]

Announcing the deal

An agreement in principle, in the form of a Memorandum of Understanding, was made on the evening of 27 August 1992, subject to Māori ratification, and was signed by the Māori negotiators and the Crown. Next morning the *Dominion* newspaper headline about the agreement indicated that the Crown would fund Māori tribes to buy half of New Zealand's biggest fishing company, Sealord Products. Carter Holt Harvey expected to raise up to $375 million by floating the company, so the deal would be for about half that amount. The joint venture company to bid with Māori was Brierley Investments. The Māori Fisheries Commission would also receive 20 percent of the quota for all fish brought into the Quota Management System from now on. The deal would give Māori control of about 40 percent of the $1.2 billion fishing industry. It was estimated that the government contribution would be up to $190 million. In return, Māori commercial fishing claims would be settled and the 1989 High Court proceedings dropped, so that the Quota Management System could recommence. Sir Graham Latimer said he expected the deal to be endorsed by tribes, despite it only delivering about half of the 50 percent of total fishing quota expected by Māori. Jim Bolger said that it had been clear, since the Waitangi Tribunal Ngai Tahu Fisheries Report, that historical grievances must be addressed.[10]

Doug Graham described the agreement as 'an expression of goodwill . . . that augers well for hearings of the future'. Tipene O'Regan said it was 'politically wholesome for the nation'. The need for the settlement was paramount because the interim agreement made in 1989 was due to expire in a month (in October 1992). Almost immediately after the deal was announced, National Party backbench MPs began questioning it, as did representatives of Ngāti Whātua in Auckland.[11] For example, Hobson MP Ross Meurant lodged a complaint with the Human Rights Commissioner over comments made on a television news programme by Maori Fisheries Commission chair Tipene O'Regan. O'Regan is reported to have indicated that if his tribe, Ngāi Tahu, did not get rights to the South Island fishery resource, 'violence would be hard to avoid'.[12]

Prime Minister Jim Bolger assured people that the government was not giving an open cheque to Māori to purchase their half share of Sealord. In response to a possible bidding war between rival groups to buy Sealord, Bolger said that if the price went above what the government had agreed with Māori, then Māori would have to fund the difference. Bolger added that the government was aware of another bid, and that was the reason

"'S Funny . . . It feels a lot lighter than it looked in the shop window!' A cartoon comments on the perceived division of benefits from the Sealord deal: Tipene O'Regan representing Ngāi Tahu with the majority and Northern Māori with their much smaller 'head of the fish' share.

ATL: H-044-003. The Dominion, 22 February 1993. Cartoonist: Frank Greenall.

they moved speedily to conclude the agreement with Māori. He believed that most people saw the deal as fair and honourable. Tainui were already expressing support for the bid.[13] In the north, however, tribes were 'lukewarm' towards the deal negotiated. Dick Dargaville said that dividends from the new company could be used to purchase more quota for Māori, pushing their total up to the 50 percent first mooted back in 1988. Matiu Rata was warning northern leaders that if the other bidder for Sealord was successful, the quota owned by the company would then be owned by overseas interests. Tipene O'Regan was saying that even though Māori were giving up commercial fishing rights in this deal, they were not giving up traditional and customary rights, even though some Māori had been saying this.[14] Even former Prime Minister David Lange had things to say about the deal. He was critical of Jim Bolger, sceptical of the deal's success, and stated that it was for the settlement of all Māori fishing claims, probably fuelling public hysteria, when in fact it applied only to commercial fishing claims.[15]

Getting Māori agreement

Despite the 'agreement in principle', the deal still had to be ratified by at least a majority of tribes and the Māori–Brierley joint venture had to win the tender from Carter Holt Harvey (CHH). The spokesperson for CHH, Wilson Whineray, said that a public float of Sealord was still possible, despite Māori and Brierley expressing interest: 'Carter will consider all bids', he said. A rival bid was expected from Ashlar Corporation, composed of Danish fishing company Royal Greenland, investment banker Ord Minnett, and former Fletcher Fishing people.[16] To bolster their forces, the Māori negotiators sought additional support: Professor Whatarangi Winiata, Maanu Paul, Dick Dargaville and David Higgins of Ngāi Tahu were added as negotiators.

Time was critical, and the Māori negotiators hastily arranged meetings throughout the country. They had about four weeks to gain national Māori approval. Sir Graham Latimer recalls that Matiu Rata attended most of the hui, while Tipene O'Regan, Bob Mahuta and himself attended as many as they could.[17] Tipene O'Regan, Shane Jones and Doug Kidd worked intensely on trying to define and encapsulate customary and traditional fisheries in the proposed legislation.[18]

The negotiators faced immediate opposition. At a hui in Northland on 13 September, Māori there refused to accept that the Waitangi Tribunal would no longer be able to hear commercial fishing claims, and were strongly against giving up Treaty rights in return for money. Shane Jones attended that hui and took verbatim notes from which the account given here (see boxed text) is drawn.[19]

Shane Jones: Recollections of Northland hui

The Māori Fisheries Hui held at Waitangi on 13 September 1992 epitomised the style of cut and thrust debate of many of the hui held in that short month between initialling the deal and signing up to the agreement. The Māori Fisheries negotiators were convinced that the deal was the best possible, while opponents argued against 'selling' Treaty rights.

Tipene O'Regan and Maanu Paul were accompanied by Amster Reedy on a charter flight to Kerikeri. They were taken onto the Waitangi marae by Dick Dargaville. There was a very large gathering of mainly Ngā Puhi and a group from Muriwhenua, including Dame Mira Szaszy, Roera Ihaka, Geranium Kapa, Paihere Brown, Arthur Kapa, Errol Murray, Makari Matiu and Rima Edwards, with son Katene. Tipene O'Regan spoke first on the merits of the deal. Bob Mahuta spoke in Māori, but was nervous after having been challenged at an earlier hui in Whāngārei as to why Tainui was speaking in Ngā Puhi about the Treaty of Waitangi. Sir Graham Latimer was his 'usual mixture of "native" cunning and National party political intelligence . . . managing to evoke the memories of Waata Tepania [from Te Aupouri] and the Ninety Mile Beach case [the legal challenge to ownership of Te Oneroa a Tohe or the Ninety Mile Beach] with liberal references to God!' Maanu Paul also spoke in Māori: 'E kīia ana he tangata horo au ki te riri. Heoi i te rā nei e Ngā Puhi e kore au e horo ki te aha, nā te mea ehara koutou i te hoariri nāku! [Translation: They say that I am someone who is quick to anger. However, today I won't be doing anything quickly. You are not my enemy.]

Dover Samuels and Manu Wynyard, Waitangi marae chairman, would not support the Deal, saying it devalued their Treaty rights. Graham Rankin, the senior kaumātua of Ngā Puhi, also spoke: 'E Tipene, tēnā koutou ko tō iwi e noho mai ki tērā motu. Kei ahau, kei a Ngā Puhi te mauri o te Tiriti. He mea whakairo te Tiriti i konei ā ka taria haeretia ki ngā iwi. Ka tae rawa atu ki a koutou ki a Ngāi Tahu kua waimāoringia te Tiriti!! Kei a Ngā Puhi te aka matua ā maku a ia e korero. Tō matou whakamāoritanga mō te Tiriti, he taonga motuhake! Tōna rite kei te Kawenata. Whakamāramatia mai ki au he pehea e taea ai te hoko i te Kawenata. E kore a Ngā Puhi e whakaae kia riro te Tiriti i a koutou hei whāwhā mā koutou hei mahi moni.'! Pūkanakana tonu a Graham, ora mai i a ia ngā kaumātua o Ngā Puhi. Tautokotia e Ngāti Wai, nā Witi McMath i kōrero, ko te tū wehe kē a Ngāti Wai!

[Translation: Tipene, greetings to you and your people who live in the other island. The mauri of the Treaty is held by me, by Ngā Puhi. The Treaty was fashioned here and then it was taken around to the different tribes. By the time it got to you in Ngāi Tahu its essence had been diluted! Ngā Puhi still holds the core of the Treaty. I will speak for Ngā Puhi. Our understanding of the Treaty is that it is a unique and special document! To be likened to a Covenant. You explain to me how one is able to sell a Covenant. Ngā Puhi will never agree for you to use the Treaty for money-making purposes. Graham visibly expressed his feelings, and in doing so brought the many elders of Ngā Puhi to life. Ngāti Wai voiced support for Graham Rankin. Witi McMath spoke for them; that Ngāti Wai intended to stand aside.]

The negotiators expressed disappointment at the Ngā Puhi rejection of the deal, and Mira Szaszy quietly asked Rima Edwards to stand and state that Muriwhenua would not be bound by Ngā Puhi and would support the deal with conditions. Shane recorded that the atmosphere was 'highly charged', particularly as Graham Rankin was a large man with a very imposing demeanour:

> Rima delivered an extraordinary whaikōrero: 'Ka timata ia ki te karakia mō Te Atua i te Pō, mai i a Papahurihia. Ka tuku i nga whakapapa mai i a Rangi rāua ko Papa ā heke rawa mai ki a Aperahama Taonui. Ka puta i a ia ngā kupu whakarite a Aperahama Taonui. Ka mea atu ia, me whakaae tātou kia riro mā te hunga tohu e kaupare atu i ngā pūngāwerewere e ārai i a tātou.
>
> Nāna i mea atu: "E pai ana e Ngā Puhi, kia ū koutou. Heoi māku a Muriwhenua e kōrero, ka hia noa atu te roa e whakamomori ana te Hiku o te Ika kia pae mai te tohoraha ki uta. Kua oti i a Muriwhenua, me whakarewa tonu te waka nei hei hari mai i nga mau o te moana ki uta e ora ai tātou. Ki te kore koutou e whakaae e pai ana, māku mā Muriwhenua koutou e pīkau!" Horo tonu a Graham Rankin te peke. Heoi, tū tonu ana a Rima, nāna i mea atu, "E mara tō horo hoki ki te tū, pai atu me waiho mā ngā wāhine kē te waiata māku!" Whati tonu te reo a Graham, hurō pai te hui i te kata, wāwā pai te whare.'

[Translation: Rima began with a spiritual incantation from Te Atua i Te Pō, of Papahurihia religion, following on by reciting genealogical links from Rangi and Papa right down to Aperahama Taonui. [Aperahama Taonui was the visionary leader of Ngā Puhi hapū Te Popoto of Utakura in the Upper Hokianga, and a founder of the Kotahitanga movement, which evolved into the Māori parliaments of the 1890s.] He then quoted metaphorical sayings of Aperahama Taonui. He then said, let us agree that we should leave it to those who have the expertise to ward off the spiders that inhibit us. He added, It's fine, Ngā Puhi, you stick to your guns. However, I will speak for Muriwhenua. How long have we at the Tail of the Fish desperately wanted a whale to come ashore. Muriwhenua's decision is that we launch this canoe to bring the fruits of the sea ashore so that we can benefit. If you don't agree, that's fine, we the Muriwhenua will carry you. Graham Rankin was quickly on his feet. But Rima, who was still standing, turned and said to Graham, My friend, gee you're real quick to get up, but it's probably better if you leave it for the women to sing my waiata. Graham was left speechless, and the hui erupted in laughter which echoed through the house.]

Some 25 delegates to the Labour Party conference in Christchurch also rejected the clause in the agreement that would settle all commercial fisheries claims. Questions about the mandate of the fisheries negotiators were also being raised around the country.[20] Ngāti Kahungunu voted against accepting the deal, and Māori fisheries negotiators were to meet representatives of the tribe to try to change their minds. At the same time, Tipene O'Regan was threatening that Ngāi Tahu might go it alone and bid for Sealord without northern tribes. By then the fisheries negotiators had been travelling the country for two weeks since the deal was initialled and had one more week to go before reporting back to government. Other small South Island groups and a group of Chatham Island Māori expressed opposition, but Tipene O'Regan dismissed this as minor and unrepresentative.[21]

The government was very eager for the deal to be ratified, and Bolger, Kidd and Doug Graham made various public threats, warnings and exhortations to assist. Graham said that the government would be willing to proceed without 100 percent Māori support for the deal. He said, 'If the Crown believes there is sufficient consensus, it seems to me that the Crown and Maori negotiators could proceed.' He felt that tribes left out could ask the Waitangi Tribunal whether the Crown had acted honourably or not; he added, 'And if they can't do it [agree] on this one [the deal] I wonder if they ever will'.[22] Shortly after, Jim Bolger warned that opposition to the deal might impede other settlements. He said there was no other proposal on the horizon for settling commercial fisheries claims.[23]

By mid-September, Matiu Rata was confident the deal would be supported by most tribes. 'I'm not saying it's perfect or will ever get absolute support, but I'm confident that we are getting close to a decision'. He said that one of the main questions tribes were asking was how much of the Sealord quota they would get and the answer was none. 'We aren't buying it [Sealord] to split it up. It would be silly to break up a company that is doing well'.[24]

In the same newspaper report, Ngāti Toa were said to be opposed to the deal. Spokesperson Matiu Rei claimed that they were unclear about what they were getting and what they were giving up.[25] However, six northern South Island tribes had accepted the deal as it was presented to them at Waikawa marae in Picton. There, only Ngāti Apa and Rangitāne were still opposed. Their acceptance was subject to non-commercial fishing rights being protected and the Sealord shares and future fisheries quota being allocated to individual tribes.[26] Seven other tribes rejected the deal, including 'a group claiming to represent Ngati Whatua', who said they would seek a High Court injunction to block it. Tuhuru hapū from the West Coast lodged a claim with the Waitangi Tribunal seeking protection of its fishing rights.[27]

Towards the end of September, Matiu Rata said that the Memorandum of Understanding was to be reviewed by Crown and Māori negotiators, in order to take into account concerns expressed at the many hui already held, with a view to then signing it and allowing High Court proceedings to be withdrawn and a bid for Sealord made. He believed that the opposition to the deal was not enough to stop it. It would be made clearer in the revision that traditional and customary fisheries were protected. Matiu Rata was also trying to get the support of the Māori MPs, because he felt that they were responsible for the legislative future of the deal.[28]

The agreement in principle was taken to national hui and some 23 marae throughout the country. The negotiators prepared a report on those hui and this appears to have satisfied the responsible Ministers that the understanding should be formalised in a deed of settlement. The negotiating teams met again on 22 September to discuss tribal views and negotiate amendments. The main concerns were how to protect traditional and customary fishing rights and how the benefits would be shared equitably with tribes. The final negotiation session ran until 3 am and was so charged with emotion that frequent stops for karakia (prayer) had to be taken.[29]

After many long hours in negotiation, in air thick with smoke (from Doug Kidd's cigarettes, Doug Graham's cigars and Tipene O'Regan's pipe), the serious negotiations, begun in late August 1992, were ending. Doug Kidd said the talks had been intellectually, emotionally and physically draining because everyone put everything into them. He believed that it was the personal rapport which had been built between the Ministers and Māori negotiators that led to success. On reflection, Bob Mahuta was also adamant that the personal rapport built between him and Doug Graham during their Tainui settlement negotiations was what was most important to achieving success.[30]

The signing

Shane Jones, at that time working for the Prime Minister's Department, recalls the evening of 23 September 1992. Wira Gardiner, Chief Executive of the Ministry of Māori Development, had organised the final hui, bringing tribal representatives together in Wellington at very short notice. The various parties had become a bit 'lost', despite a number of large delegations having arrived to sign the finally agreed text. Changes to the original agreement had been made as a result of the consultation hui feedback, but some leaders were still unsure. The Acting Prime Minister, Don McKinnon, and other ministers signed for the Crown, together with the Māori negotiators. A few others also signed, but government officials were having difficulty gathering in enough signatures and convincing delegates. Apparently word was sent of this situation to Dame Te Atairangikaahu (the Māori Queen), who had come with Tainui to Wellington and was back at her accommodation. She hastily arranged a good number of Tainui representatives to head back from their hotel. As they made their way into the parliamentary venue, led by kaumātua Pumi Taituha, the doors burst open and in they came, karakia and waerea and signed the agreement, leading to a rush of other tribal representatives wanting to sign. Proceedings had gone on so long that representatives were hungry and threatening to leave to get food, adding to the perplexity of government officials.[31]

The weeks of high drama culminated in a Deed of Agreement being signed by most of those present. Tipene O'Regan said that the rights of his people were protected. Doug Graham said that not all tribes had signed, but the majority had, and particularly those most involved in fishing. Doug Kidd said that details of the protection of traditional fishing rights were still to be worked out, and this had been the most difficult area because of its emotional and cultural significance.[32] In the days that followed, Jim Bolger's support was required to convince several National Party MPs that it would work and they should not oppose it.

The legislation giving effect to the deal was introduced into Parliament the next day. Justice Minister Doug Graham ended his introductory speech in tears, and was given a standing ovation. Members of Parliament on all sides shook his hand in congratulation. He later said that this event was one of the most emotional moments of his life. He felt the Crown had acted honourably now, whereas in the past some Crown actions had been dishonest and wrong. He believed that the previous Labour government had raised

OPPOSITE PAGE

Wellington, on the evening of 23 September 1992: Acting Prime Minister Don McKinnon (nearest camera), Chair of the Māori Fisheries Commission, Tipene O'Regan, Minister for Treaty Settlements, Doug Graham, Maanu Paul and Fisheries Minister Doug Kidd signing the Deed of Settlement while at the far end of the table Sir Graham Latimer looks on. Hon. Matiu Rata is out of shot to the left of Don McKinnon.

Fairfax NZ: Michael Smith, 10-DPT-fishing1992 Dominion Post

awareness of Māori grievances. Negotiations were not easy, particularly for the Māori negotiators, 'who had the burden of knowing their ancestors were talking to them'.[33]

The opposition emerges

Tribes opposed to the Sealord deal began High Court action on 5 October 1992. Initially, groups from the Chatham Islands, Ngāti Whātua and Rangitāne began the action. Their principal objection was that the Māori fisheries negotiators did not have a mandate to sign the deal. However, the Prime Minister had stated that tribes not signing the deal would not be excluded from its benefits. Sharing of the benefits would be up to Māori, with assistance from the Minister of Māori Affairs, he said. He felt that the deal would be successful and that all New Zealanders should celebrate it.[34]

Margaret Mutu summed up opposition to the deal at the time in her book, *The State of Māori Rights*:[35]

> The Māori negotiators were men chosen by the Crown. They had no mandate to act on behalf of all iwi or all Māori and for some, their mandate within their own iwi or organisation was not even sure. And although the Māori negotiators called hui to consult, they disregarded significant opposition and made positive recommendations to the Crown. Negotiations were reported as being conducted with strong emotional overtones. Simple commercial considerations such as the size and value of the fishing resource, and a careful examination of the viability of either the international fishing industry or the fishing company purchased as part of the settlement deal, were accorded very little attention. Moana Jackson of the Māori Legal Service was reported as noting that instead of grabbing what was offered, the fundamental difficulties with the quota management system, the guidelines and instructions for negotiations, the disbursement of the income and the mandate of the negotiators should have been worked out first, before negotiations began.

This summary, however, is disputed and subsequent court and Waitangi Tribunal scrutiny did not support this opposition enough to halt the deal.

Tribes also lodged a claim with the Waitangi Tribunal. Muaupoko, which has claims to commercial fishing areas around the Manawatu river mouth and coast opposed the deal and joined the Ngāi Tahu hapū, Tuhuru, in its claim to the Tribunal.[36] The original litigants were joined by more tribes, Ngāti Awa, Ngāti Kahungunu and Ngāti Porou, who claimed that they were the majority of Māori.[37] The court was also told that only sixteen out of 54 government-recognised tribes had approved the Sealord deal at a Māori Congress[38] hui (on 19 September), four days before the official signing. However, The Crown argued that it was inappropriate for the courts to rule on the deed of settlement because it was basically a matter of public policy, and the courts should not interfere.[39] Justice Heron of the High Court agreed, but some of the tribal opponents of the deal said they would appeal. Maui Solomon, spokesperson for one group, said the groups would continue to challenge the deal.[40]

A week later, opposed Māori groups, including elements of Ngāti Whātua, Ngāti Awa, Rangitāne o Wairau and Wharekauri Rēkohu, went before the Court of Appeal to try to stop the government passing legislation giving effect to the Sealord deal. Sir Robin Cooke, President of the Court, commented that the Sealord deal could be seen as a reasonable and equitable method of solving a complex and difficult problem. He said that for the objectors to succeed, they might have to show that the government was acting unreasonably or in breach of its fiduciary duty under the Treaty.

The argument for those opposing the deal, said their lawyer, was that they wanted 'fishing quota rights, not paper in a fishing company'. He added that there was no agreement on distribution of the benefits, and that the negotiators had assumed majority support for the Deal when in fact none of the tribes had actually seen the deed of settlement

ABOVE

Dr Margaret Mutu of Ngāti Kahu with Ngāti Kahu kaumātua Tuhoe Manuera at Maitai Bay, 2004. Margaret was a critic of the 1992 Sealord Deal. However, in 1993 she became a member of the joint Crown–Māori working party convened to draw up customary fisheries regulations.

TOK

RIGHT

The president of the Court of Appeal, Sir Robin Cooke (right) and the third-ranking judge, Mr Justice McMullin, walk back to the Court from Parliament Buildings, 20 October 1986. Sir Robin had just been sworn in by Mr McMullin as the Administrator of Government. Robin Cooke and the Court of Appeal have been hugely influential in the Māori fisheries story.

Newspix.co.nz: 1033256 Paul Estcourt, NZ Herald

until the day it was signed.[41] The opponents did not want to be bound by the deal. However, the Crown countered this by arguing that the Māori leaders who signed the deal did not claim to be representing all Māori.

Sir Robin Cooke commented during proceedings that passing a law to block fishing claims already before the Waitangi Tribunal would be an extreme step. He also noted that it had not been decided how widely the Treaty of Waitangi could be enforced. He said the orthodox view was that the Treaty was not enforceable except where it was incorporated in Acts of Parliament.[42]

On 3 November, the Court of Appeal judgement cleared the way for the Sealord deal to go ahead. The judgement stated:

> The proposal of the Crown and the Maori negotiators to endeavour to obtain a substantial Maori interest in Sealord is thoroughly consistent with the approach of this Court in previous cases . . . The Sealord opportunity was a tide which had to be taken at the flood . . . a responsible and major step forward has been taken.[43]

Sir Robin Cooke said that the court would not overturn the previous High Court decision. He said that the deal was a responsible and important step forward, and Māori failure to take it up might be seen as a failure of duty by a partner in a partnership. The deal had been negotiated by responsible Māori leaders and Parliament must be free to decide the law.

'All that can safely be said is that the Deed was negotiated by some responsible Māori leaders and has significant Māori support but also significant Māori opposition.'[44] The deal was a political compact; it did not repeal the Treaty of Waitangi and was binding only on the signatories. The Court was not concerned with the wisdom of the laws or the degree of Māori mandate – these were 'political questions for political judgement'. Doug Kidd said the court decision was 'wonderful news for Maori, the fishing industry and for New Zealand'.[45]

On 6 November 1992, The Waitangi Tribunal made public its report on the claims by opponents of the Sealord deal. The Tribunal found fault with the deal, but did not uphold most of the claims against it. It suggested ways in which the deal could be made fairer and maintain Treaty rights. It also agreed that the Māori fisheries negotiators had consulted widely and achieved consensus among Māori for the deal to go ahead.[46] The main feature was that it agreed there was a mandate for the deal.[47] The Tribunal stated, 'the settlement has rightly been hailed as historic'.[48]

Māori buy Sealord

A week after the Tribunal report, Carter Holt Harvey (CHH) announced the sale of Nelson-based Sealord Products to the Māori Fisheries Commission/Brierley Investments joint venture. The company owned 25 percent of all fish quota and had achieved $247 million in sales in the previous year. It employed about 1400 people. The Māori Fisheries Commission would thereby control about one third of the New Zealand fishing industry.

Prime Minister Jim Bolger said that the sale concluded the Crown's obligations to Māori under the Treaty of Waitangi for commercial fishing rights. Doug Kidd, Minister of Fisheries, said that legislation would be introduced soon.[49] It was reported that an appraisal report done for CHH showed that the Māori/Brierley bid returned several million dollars more than the rival bid.[50]

Fishing industry leaders were alarmed at law changes proposed to give effect to the Māori fisheries deal. They claimed that the government planned to privatise the industry. Fishing Industry Association president Eric Barratt said that the government was planning to sell new fish quota to the highest bidder, rather than allocate it on the basis of fishers' catch histories. This was a breach of Fisheries Minister Doug Kidd's previous undertaking. All commercial fisheries could be brought into the QMS now that the Māori fisheries

Sealord products; Foreign interests. This cartoon shows a small waka, in which Jim Bolger sits with a net ready to scoop up the fish (Sealord Products) that Matiu Rata has hooked on his line. Down below a shark (foreign interests) is eyeing the proceedings. This refers to the government's decision to put up money to enable Māori tribes to buy half of Sealord Products (New Zealand's largest fishing company) to settle Treaty of Waitangi fisheries claims.

ATL: H-098-001. NZ Herald, 1 September 1992. Cartoonist: Laurence Clark [Klarc]

RV *Tangaroa*, the research vessel operated by the New Zealand National Institute of Water and Atmospheric Research, c. 1992. *Tangaroa* conducts fisheries and marine research around New Zealand. It is equipped for hydrographic, bathymetric and oceanographic surveys to measure and map various properties of the ocean and seabed; biological surveys; and for both acoustic and trawl fisheries surveys. It can trawl to a depth of 4,000 metres. On one voyage in 2003, scientists aboard *Tangaroa* discovered over 500 new species of fish and 1300 species of invertebrates.

TOK

issue was resolved. Barratt felt that some existing operators might not be able to afford to buy enough quota to keep their businesses profitable. He said also that the government's plan to increase (roughly by double) resource rentals in the next few years was to cover the cost of running the fisheries division of government. So the government was giving to Māori (by quota) with one hand, and taking away (by higher resource rentals) with the other.[51]

Opposition also came from within the government's own ranks. Four government MPs voted with the opposition to oppose the legislation giving effect to the Sealord deal, the Treaty of Waitangi (Fisheries Claims) Settlement Bill. The dissidents expressed anger that the Bill did not go through the normal Select Committee process, and claimed that backbench MPs had not been consulted about the Bill's final content. The Bill allowed the Fisheries Minister to recognise and provide for traditional food gathering (rights) by Māori and to involve Māori in managing those rights. Mātaitai (food gathering reserves) would be able to be set up, and opponents said they feared that most New Zealanders would be excluded from them. One opposing MP called Doug Kidd and Doug Graham puppets, and claimed they had acted without the agreement of the Caucus. The Deal was said to reek of racism and favouritism toward Māori and the Bill was divisive.[52]

All six Māori MPs opposed the Sealord legislation. Koro Wetere said it was a mistake to abolish Treaty fishing rights. Peter Tapsell claimed that Māori on tribal fisheries committees did not have sufficient management skills.[53] Winston Peters criticised the legislation not being sent to a Select Committee for discussion, and accused the government of hypocrisy and arrogance. He raised again the claim that the Māori negotiators, other than Tipene O'Regan (who chaired the Ngai Tahu Maori Trust Board), had no formal mandate.[54]

In a separate article on the front page of the *Dominion*, Jim Bolger moved to allay public fears about exclusive Māori fishing areas. He said these were not new and would be small areas of the coast, mainly shellfish gathering areas.[55] In an emotional speech in Parliament, Doug Graham called on all New Zealanders to support the Sealord deal (his speech was interrupted several times by Winston Peters). Graham also defended the way the Bill was rushed through Parliament. There was no point in inviting public input when changes could not be accepted, as the agreement had been settled, he said.[56]

The Treaty of Waitangi (Fisheries Claims) Settlement Act 1992 was passed by Parliament under urgency on 10 December 1992. The Act provided the legislative basis for the full and final settlement of Māori claims to commercial fishing rights, namely: the Crown's funding of a Māori joint venture acquisition of Sealord; the establishment of the Treaty of Waitangi Fisheries Commission, and the appointment of up to thirteen Commissioners; the repeal of statutory recognition of Māori fishing rights; the limitation of the Waitangi Tribunal's powers of inquiry into commercial fisheries claims (but not the extinguishment of Treaty rights); provision for the recognition of Māori non-commercial (customary) fishing rights; and provision for the formulation by the new Commission of accountability measures, and a system for the allocation of the benefits of Sealord (both arrangements to be recognised in legislation once they were agreed by Māori). Legislation providing for the transfer of 20 percent of new quota to the Commission was also signalled.

Prime Minister Bolger criticised those who opposed the deal, claiming that 'they don't seem to have it within them – neither their hearts nor their minds are big enough to see what we have done'.[57] However, opposition to the legislation continued. The Māori Congress had instructed former Māori Affairs department secretary Dr Tamati Reedy to file a case with the United Nations protesting against the deal.[58] In a letter to the editor of the *Dominion*, Dick Dargaville (who represented Auckland on the National Party Maori

Advisory Board) criticised the trip by Dr Reedy to the United Nations, where he criticised the Sealord deal and the role of the Māori negotiators. Dargaville pointed out that Reedy was head of the Department of Māori Affairs when the QMS was first introduced, and felt that the criticism by the Māori Congress at the United Nations was potentially divisive among Māori.[59]

The New Treaty of Waitangi Fisheries Commission

Māori Affairs minister Doug Kidd announced that he would call a hui in early 1993 to discuss criteria for selecting members of the new Treaty of Waitangi Fisheries Commission, which, under the new legislation, replaced the Māori Fisheries Commission. He had appointed five interim commissioners who would hold those positions until 26 February 1993, when thirteen permanent appointments would be made. The five interim Commissioners were Tipene O'Regan, Dame Mira Szaszy, Whaimutu Dewes, Philip Pryke and Nick Jarman. The new commissioners would begin work on how the new Commission would be accountable to Māori.[60]

The tumultuous year ended on a high point. Māori fisheries claims were settled as far as the Crown was concerned. The fishing industry had been given certainty, although resource rentals for quota were still creating anxiety and controversy. Māori now had a huge opportunity to take what they had won and grow it into something fantastic. The future of Māori fisheries was now essentially in their own hands.

On 6 January 1993 the Māori/Brierley Investments joint venture took possession of Sealord. On 16 February Doug Kidd, in his role as Minister of Māori Affairs, convened a national hui to discuss the appointment of Fisheries Commission members, and in May 1993 they were announced: Tipene O'Regan (Chair), Sir Graham Latimer (deputy Chair), Hon. Ben Couch (former Minister of Māori Affairs), Whaimutu Dewes, Craig Ellison, Shane Jones, Robert Mahuta, Dr John Mitchell, Naida Pou, Phillip Pryke, Anaru Rangiheuea, Archie Taiaroa and Evelyn Tuuta.

ENDNOTES

1. Doug Kidd, interview with Brian Bargh, 2014.
2. Ibid.
3. Robert (Bob) Mahuta, interview with Adam Gifford, September 1995.
4. Ibid.
5. Tipene O'Regan, interview with Brian Bargh, April 2015.
6. Robert (Bob) Mahuta, interview with Adam Gifford, September 1995.
7. Graham Latimer, interview with Adam Gifford, 1995.
8. Ibid.
9. Ibid.
10. *Dominion*, 28 August 1992. p. 1.

11. *Dominion,* 31 August 1992. p. 2.
12. *Dominion,* 13 August 1992. p. 2.
13. *Dominion,* 1 September 1992.
14. *Dominion,* 3 September 1992. p. 8.
15. *Dominion,* 7 September 1992. p. 8.
16. *Dominion,* 1 September 1992. p. 15.
17. Graham Latimer, interview with Adam Gifford, 1995.
18. Tipene O'Regan, pers. comm., 2014.
19. Shane Jones, pers. comm., 2015 (based on notes taken 13 September 1992 at Waitangi).
20. *Dominion*, 9 September 1992.
21. *Dominion,* 11 September 1992. p. 1.
22. *Dominion*, 12 September 1992.
23. *Dominion,* 15 September 1992. p. 2.
24. *Dominion,* 14 September 1992. p. 1.
25. Ibid.
26. *Dominion,* 17 September 1992. p. 13.
27. *Dominion,* 18 September 1992. p. 3.
28. *Dominion,* 22 September 1992. p. 7.
29. *Dominion,* 6 October 1992. p. 7.
30. Robert (Bob) Mahuta, interview with Adam Gifford, September, 1995.
31. Hon Shane Jones, pers. comm., March 2015.
32. *Dominion,* 24 September 1992.
33. *Dominion*, 25 September 1992. p. 2.
34. *Dominion,* 6 October 1992. p. 1.
35. Margaret Mutu, *The State of Māori Rights*. Huia Publishers, 2011. p. 14.
36. *Dominion,* 26 September 1992.
37. *Dominion*, 8 October 1992.
38. The National Māori Congress was founded on 14 July 1990 and represented about 37 tribes. It was independent of government and had a mandate to represent Māori views on issues relevant to all Māori.
39. *Dominion*, 10 October 1992. p. 3.
40. *Dominion,* 13 October 1992. p. 13.
41. *Dominion,* 20 October 1992. p. 3.

42. *Dominion*, 21 October 1992. p. 3; 22 October 1992. p. 8.

43. *Te Runanga o Wharekauri Rekohu Inc and ors v Attorney-General and ors* (CA 297/92, judgement 3.11.92). pp. 12, 13, 18.

44. *Tangaroa,* No. 12, December 1992. p. 3.

45. *Dominion*, 4 November 1992. p. 1.

46. *Dominion*, 6 November 1992. pp. 1, 3.

47. *Dominion*, 7 November 1992. p. 2.

48. Waitangi Tribunal, *Fisheries Settlement Report*. Department of Justice, 1992. p. 6.

49. *Dominion*, 18 November 1992. p. 1.

50. Ibid, p. 17.

51. *Dominion,* 19 November 1992. p. 2.

52. *Dominion,* 5 December 1992. p. 2.

53. *Dominion*, 8 December 1992. p. 8.

54. Ibid, p. 2.

55. Ibid, p. 1.

56. *Dominion*, 9 December 1992. p. 2.

57. *Dominion*, 11 December 1992. pp. 1, 2.

58. *Dominion*, 19 November 1992. p. 2.

59. *Dominion*, 24 December 1992. p. 6.

60. *Dominion,* 26 December 1992. p. 1.

PROCESSING THE CATCH:
Expanding Māori fisheries assets

The events leading up to the settlement of Māori commercial fisheries claims by the purchase of Sealord were historic by anyone's reckoning. The new Treaty of Waitangi Fisheries Commission (also known as Te Ohu Kaimoana or TOK) began work as soon as it was appointed in May 1993, building on the preparatory work done by the previous Maori Fisheries Commission and the interim Commission appointed by Doug Kidd. There was still a reasonable level of opposition to the Sealord deal from Māori constituents, much uncertainty about what it would mean in practice, and a high level of expectation from the various signatories to the Deed.

On top of the pre-settlement assets, Māori now owned half of New Zealand's biggest fishing company — Sealord. They also received 20 percent of the commercial quota of any new fish species brought into the QMS. The new Commission was charged with promoting Māori commercial fishing, making sure that it was more accountable to Māori and ensuring that it had greater influence in fisheries management.

More specifically, the Commission needed to work on: how to protect Māori customary fishing rights — to be covered by regulations; develop a procedure to determine who would benefit from the settlement; and agree on a scheme for the distribution of benefits. Once those matters were agreed among Māori, new legislation to give effect to them would be enacted.

Within the wider New Zealand fishing industry, following the settlement of Māori fisheries, there was still uncertainty about quota, resource rentals and paying for necessary research and control. The government needed to make changes to the total allowable catches in the light of better research. This could necessitate changes to Individual Transferrable Quotas and fishers were worried. If quotas were to be reduced, would they be compensated? The government decided to freeze resource rentals until September 1994, and keep these in a fund to pay compensation to fishers for possible cuts to quota. David Johnson, in his book *Hooked: The Story of the NZ Fishing Industry,* summarised the situation of the industry as whole at that time:

> The Maori settlement was a major milestone. Maori controlled fishing companies now joined the Fishing Industry Association. Communication throughout the industry was better than it had ever been . . . The participants in the industry were as rugged and as uncompromising as they ever had been.[1]

The New Zealand fishery in 1990

The Waitangi Tribunal had been given a sketch of the New Zealand fishery during the earlier hearings into the Ngāi Tahu claim. The Tribunal was told that the fishery stretched over more than 30 degrees of latitude in the southern Pacific ocean, and that New Zealand is surrounded by one of the world's largest fishing zones, the 200 mile Exclusive Economic Zone (EEZ). This area covers more than 3 million hectares and contains a substantial range of fisheries resources. The modern commercial fishing industry is based on approximately 80 species out of the many thousands which occur

New Treaty of Waitangi Fisheries Commissioners appointed by the Minister of Māori Affairs, Parekura Horomia, Wellington, September 2000. Top row, from left: Hon. Koro Wetere, Craig Ellison, Toro Waaka, Maui Solomon, Robert McLeod, Archie Taiaroa, (CEO) Robin Hapi. Front row, from left: June Mariu, Naida Glavish, Shane Jones, June Jackson. Absent: Ken Mason

TOK

in the EEZ, and operates in two distinct zones, inshore and offshore fisheries.

The inshore fisheries lie on the continental shelf, reaching out to a depth of about 200 metres. The offshore fisheries occur along the continental slope, beginning at 200 metres depth and extending to 1000 metres. Again, there is great variation in the distance from the shore for this type of fishery.[2] In 1990, total catches were very large:

> . . . more than 578,000 tonnes of fish and shellfish were caught and earnings from this catch amounted to approximately $1 billion, with nearly $750 million of that coming from export receipts. More than 8000 people are employed in the catching and processing sectors of the industry and many hundreds of millions of dollars are currently invested in quota holdings and fishing plants. Commercial fisheries have developed into one of New Zealand's leading export industries and they are managed under one of the most advanced, albeit controversial, management regimes in the world.[3]

Measuring progress

In September 1993, TOK presented its first annual report to the Minister of Māori Affairs, who by that time was John Luxton. The report set out TOK's main functions: to assist Māori entry into the fishing industry; to establish a public company called Aotearoa Fisheries Ltd (AFL) and hold all the shares in that company; and to transfer to AFL at least 50 percent of the quota and cash paid to TOK by the Crown (as agreed in the settlement).[4] The original AFL was wound up after paying a 'special dividend' to TOK for it to purchase Sealord. On 24 June 1993 it was liquidated and its remaining assets were transferred to TOK.[5]

By its own account, TOK had made good progress in implementing the settlement, and all the 10 percent of Total Allowable Commercial Catch due to TOK from the Crown had been transferred by October 1992. The Crown was still to obtain and transfer back-dated quota, as had been agreed in 1990. The Commission was in the process of leasing the quota they held by tendering it, and a high proportion of the tenders were from Māori.

Work had begun on the new Maori Fisheries Act, as the settlement Act required. The key areas for development were: a scheme for the allocation of post-settlement assets; defining the functions of the new Commission; a procedure to identify beneficiaries of post-settlement assets; and a process for ensuring that TOK was accountable to Māori. A further important matter was that of Māori customary (non-commercial) fishing rights. During the 1992–93 year, TOK had been working with the Crown to develop new regulations that would protect those rights. A series of nearly 30 hui had been held throughout the country to discuss the matter.[6]

During that same year, iwi had requested funding to help develop their fishing interests and prepare for their quota allocation. However, TOK had delayed any distribution of such funding until a proper system was put in place and its cash-flow could cope with the 'considerable' requests.[7]

By October 1993 TOK had released a discussion document: *Pre-settlement Assets. Possible Methods for Allocation*.[8] This outlined six possible models for discussion. After discussion and consultation with iwi, the six models were whittled down to three by August 1994. The pre-settlement assets consisted of about 57 thousand tonnes of quota, $58 million in shares in Moana Pacific Ltd, and nearly $50 million in cash. The discussion document outlined the options for allocation: a population based formula, a formula based on a tribe's marine rohe (area of authority, mana moana), or something in between. There was also

OPPOSITE PAGE

Rough weather at sea: this Sealord trawler demonstrates the tough conditions for deep-sea trawling, June 1992.

ATL: EP-Industry-Fishing-NZ Industries

the major problem that most iwi were not ready to cope with receiving any assets. They simply had neither the expertise nor the infrastructure with which to manage them.[9]

Pressure was already being applied to TOK through a newly formed grouping calling itself the Treaty Tribes Coalition. Members of the Coalition included the Hauraki Māori Trust Board (representing the 12 iwi of Hauraki), Ngāti Kahungunu Iwi Incorporated, Ngāi Tamanuhiri Whānui Trust, and Te Rūnanga o Ngāi Tahu. The group claimed to represent 15–20 percent of the Māori population, relating to approximately 60 percent of the coastline of New Zealand.[10] Iwi submissions on the discussion document were not considered helpful, so the Commissioners drew up a new document, *Allocation of Pre-settlement Assets – Models for Allocation,* and released this in October 1994. This document was distributed widely and hui were held to gauge opinion and hear submissions on the models proposed.[11]

The chairman and executive of the Treaty Tribes Coalition when they met the Treaty of Waitangi Fisheries Commission, Wellington, 4 August 1994. Back row, from left: Greg White, David Higgins, Paul Morgan, Harry Mikaere, Tom Paku, and Witi McMath. Second row, from left: Lisa Paranihi, Tiopira Te Rauna Hape, Wayne Peters. Front: Tukekawa Wyllie (chairman). The Treaty Tribes executive had walked out of the Commission hui held the day before.

ATL: EP-1994-2395 - Melanie Burford, Evening Post

Treaty Tribes Coalition

Sir Tipene O'Regan recalls, in an interview with the author, the formation of the Treaty Tribes Coalition (TTC) and the arguments members advanced[12]:

> [TTC] came about because it was suggested to me by one of our lawyers that there needed to be a grouping of tribes who had a clear Treaty based interest in the Treaty and the question of allocation, but more generally about the fisheries sector and our whole approach needed to be more disciplined and more . . . locked in on the nature and extent of the Treaty right.
>
> And if you take the view that our greatest triumph was the one that basically Whaimutu [Dewes – also a TOK Commissioner] and I led the intellectual battle for . . . we worked out a way of articulating a Treaty right through the [fisheries] management system. What was clear was that further development needed to be consistent with the Treaty and there needed to be an advocacy group to speak for that. But as Ngāi Tahu was going to be involved in this, it needed to make sure it had a wider span, a wider block of support. The groups that were thinking like us were Hauraki and Kahungunu. Others were interested but more concerned about outcomes than they were about how. So we formed the Treaty Tribes Coalition and the decision to do so, the argument for it was put at a hui held at Bidwill Street [Sir Tipene's family home in Wellington], and the man speaking for Kahungunu was their kaumātua, John Scott. Harry Mikaere and Toko Renata [from Hauraki] and David Higgins and one or two others for Ngāi Tahu were there.
>
> I took the view that as chairman of the Fisheries Commission, I should not be directly involved. I was always conscious of it. We saw ourselves as a counter to some degree to the aspirations of the other group, which wasn't led so much by Shane Jones at that stage but by various other elements who were based around the population thesis. The population thesis was simply [that] population became a proxy for need. And so essentially the whole argument developed into an argument between rights and needs, and the proxy for need became population or location. The rights thing was tied to whakapapa and place. We developed an argument in the courts, and developed an argument vis a vis the Crown [argument] based on rights and location and needs, and the counter to that was this equity distribution model. So essentially it was a group to do that, but the argument for continuance of the Treaty Tribes came with the final allocation arrangements, that there was going to be a time for a review. That was in the Act. So it needed to be kept alive and breathing and working in order to prepare Māoridom for the review. Our successors have not been particularly diligent about preparing for that.

Customary and traditional fisheries continued to be tackled by TOK. After extensive consultation with iwi throughout the country to discuss the 'nature and extent of customary and traditional fishing rights', a discussion document, *Mahinga Kaimoana Tuturu,* was released, and submissions to TOK were used to help draw up a set of customary fisheries regulations. As a result, the Minister of Fisheries asked that a joint working party of four Crown and four iwi representatives be convened. Guided by further hui and consultation, the iwi representatives appointed were Dr Margaret Mutu, Rakiihia Tau, Maui Solomon and Caren Wickliffe. They had very specific instructions to ensure, among other things, that tino rangatiratanga over non-commercial customary and traditional fishing rights was upheld in the regulations and legislation that might ensue.[13]

In the meantime, the Commission reported that while the allocation of assets was still being considered, it had continued to make fish quota available to iwi through leasing arrangements. The Commission agreed to lease this quota to iwi at slightly less than commercial rates. This protected and enhanced the value of assets under Commission control.[14] The Commission had already set in place a significant training and development programme, with the aim of facilitating Māori into the business of fishing – as was their mandate.[15]

So the second year of existence was one of consolidation for TOK. Their annual report to the Minister of Māori Affairs, John Luxton, was crammed with exciting and challenging tasks. The future of Māori fishing seemed to be in good hands. The year was dominated by concerns that TOK be accountable to iwi, and their decisions and actions seemed to reflect that imperative. Tipene O'Regan became a Knight in the Queen's birthday honours – another positive for Māori.

However, two major issues were clearly the focus of significant effort by TOK:

- The need to protect and enhance its fisheries assets, and
- The requirement to allocate assets to tribes.

These two major themes were to take thousands of hours and millions of dollars worth of effort to resolve.

Other activities were important adjuncts to these two critical issues. The work of TOK divided roughly into four streams of activity. First, allocation of assets: whether to allocate or not, and if so, on what basis and to whom, was the main focus of attention. Secondly, protecting and enhancing the fisheries assets. Thirdly, developing effective rules and regulations around Māori customary and traditional fisheries, while taking account of the concerns and prejudices of the general public. Fourthly – perhaps simpler, but no less critical – was the requirement on TOK to facilitate Māori into the business of fishing.

Allocation of assets: the dispute drags on

In 1995 litigation continued in the High Court over allocation of TOK assets. The Treaty Tribes Coalition had lodged claims against TOK. As well, four urban Māori organisations had challenged the TOK. The court was asked to rule on whether Māori urban authorities, representing thousands of Māori living in cities, should be allocated some of the fisheries assets. In June, Judge Anderson ruled that a preliminary question had to be determined: 'is the Commission required to allocate pre-settlement assets solely to iwi?'

TOK itself had gone to court to challenge the Waitangi Tribunal's ruling that the Tribunal had the right to hear claims made by Māori against it. In the High Court, Justice Ellis ruled on 1 August 1995 that, although the Commission's action was premature (since an allocation model had not yet

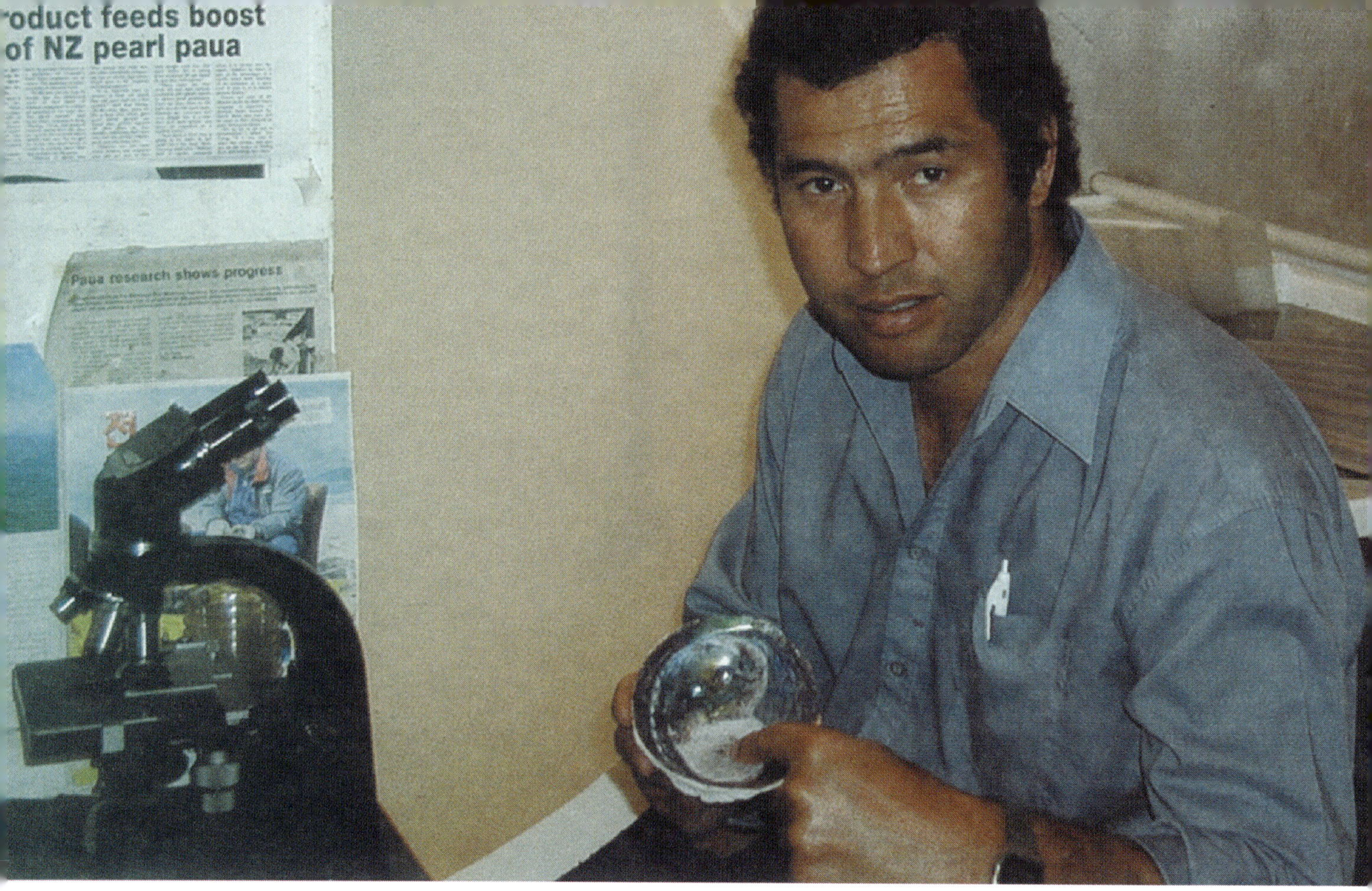
feeds boost
of NZ pearl paua
Paua research shows progress

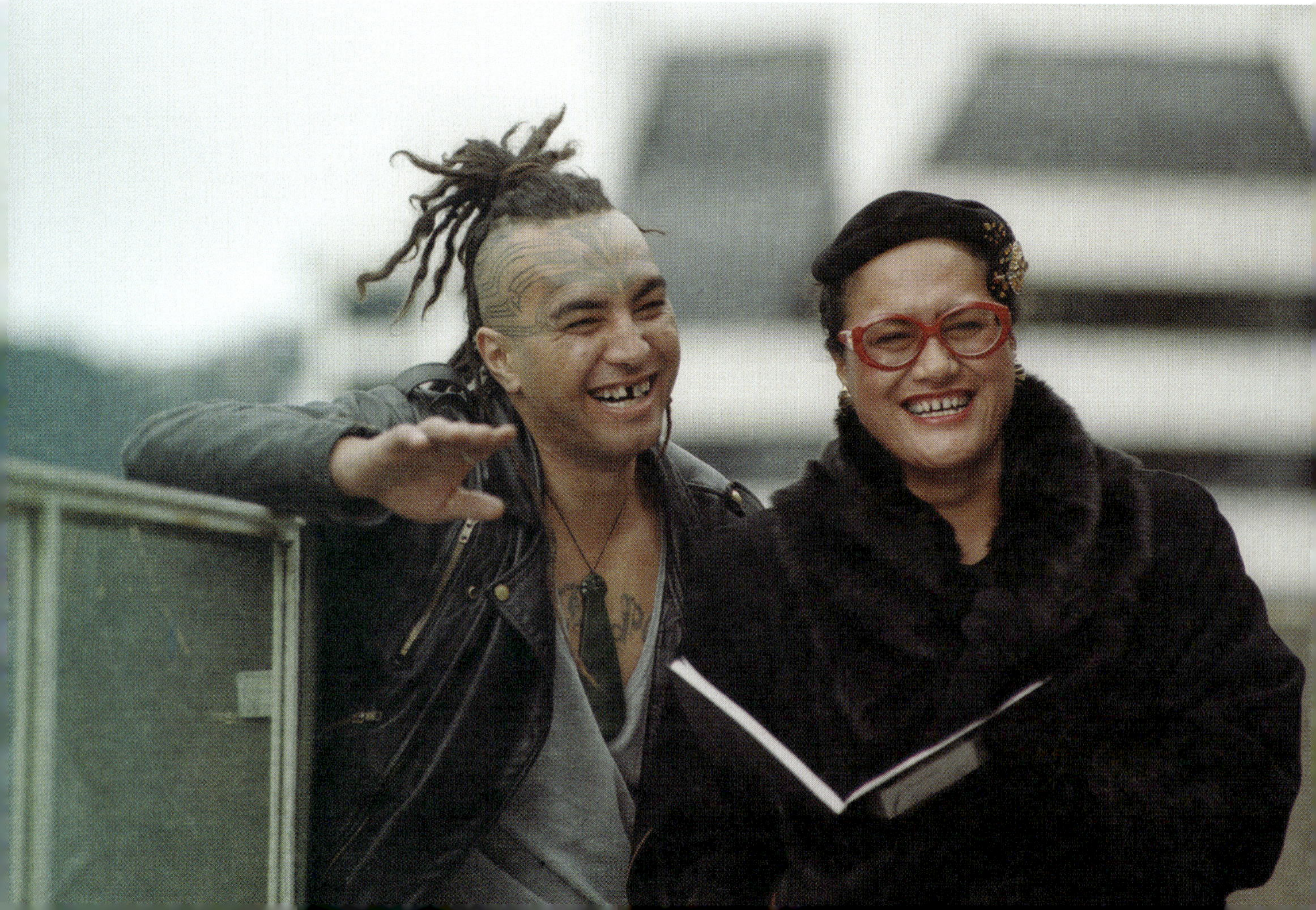

been agreed), the Waitangi Tribunal was entitled to hear claims against TOK. This ruling was appealed by all parties, and the hearing was due in October 1995.[16]

By July 1995, the allocation question was continuing to make little progress, due in part to the continuing court action. At the TOK's annual meeting (Hui-ā-Tau), a working party of iwi representatives (to be called Taumata Paepae) was charged with helping to develop an allocation model. At the meeting the 'Area One Consortium' (a group favouring allocating fisheries assets based on both length of coastline and population) and the Treaty Tribes Coalition (who favoured a model based mainly on a tribe's length of coastline) agreed to sign an agreement to call off all court action and work together alongside TOK to come up with an agreed method of allocation. The groups agreed that both methods of allocation would be incorporated in any final agreed method.[17] January 1996 was set as a deadline for coming up with a model for allocation. Members of the Area One group included Ngā Puhi, Te Whānau ā Apanui, Muriwhenua, Maniapoto and Ngāiterangi.[18]

After further consultation, TOK agreed to broad principles within the agreed framework. For example, allocation would be to iwi only; inshore and offshore fisheries needed to be defined; inshore fisheries would be allocated on the basis of mana moana–mana whenua, determined by authority over length of coastline; offshore fisheries would be allocated on the basis of both mana moana–mana whenua and iwi population; and cash and shares would be allocated in proportion to the relative quota tonnage.

In a move to try to prevent disputes and court action, TOK established a disputes resolution process.[19] It was hoped to avoid costly court action by first trying to reach a mediated solution. In April 1996, the Court of Appeal released its decision on Māori urban authorities, ruling that any allocation scheme should include a provision for them and that TOK should consult with them to seek a method of ensuring that they were able to benefit from the fisheries settlement. This decision was appealed to the Privy Council, and the hearing was due in November 1996.

In July Taumata Paepae reached agreement on a series of elements for the allocation model, including that the pre-settlement assets should be allocated to iwi; and in December 1996 TOK released a document that defined minimum standards which iwi organisations would have to achieve in order to be 'recognised' and be able to receive fisheries asset allocation (called 'preparedness').[20]

The new Minister of Māori Affairs, Tau Henare, received TOK's 1997 annual report from its Chair, Sir Tipene O'Regan. The number of commissioners remained at twelve, following the death of Ben Couch the previous year. In January 1997, the Privy Council upheld TOK's appeal and quashed the Court of Appeal finding that urban Māori needed to be included in any allocation scheme.

The allocation model was still being argued through, and in July 1997 TOK released a document,

OPPOSITE PAGE TOP

Quentin Goldsmith, c. 2004 at a disused chicken hatchery near Gisborne where he is experimenting with growing seafood in tanks. He received a grant from TOK to study aquaculture at Deakin University in Australia and was planning to set up a major aquaculture business.

TOK

OPPOSITE PAGE BOTTOM

17 June 1997: Eugene Ryder, one of four urban Māori, with their lawyer Donna Hall outside court, who fought for urban Māori to share in the benefits of the Māori fisheries settlement. Ryder later withdrew from the case saying he had achieved his goal of drawing attention to the plight of urban Māori. Known at the time as the 'Ryder proceeding' (CP 171/97) the case was against TOK's decision to allocate fisheries settlement assets to iwi defined as 'traditional tribes'.

ATL: EP-1997-1709

Proposed Optimum Method for Allocation, for the annual Hui-ā-Tau to consider. The meeting was attended by about 1000 people. There was 'general consensus that [the model] should go ahead, and be based on a compromise between population and coastline'.[21] During September and October, fifteen consultation hui were held with iwi to discuss the 'optimum' allocation method. However, several Māori organisations continued their court challenges and opposed this model.[22]

Whatever spin TOK chose to create, this 1997 year saw little further progress towards allocation, a seemingly endless amount of court action, and another year of frustration for iwi and Commissioners alike. Jenny Shipley, feeling frustration of another kind with Jim Bolger's leadership of the government, chose to lead a 'coup' in December 1997. Bolger was dumped and she became Prime Minister. However, perhaps the most tragic news had come on 18 July, when Matiu Rata died in a car crash. He was 63. Tributes to Matiu flowed from many quarters.

In February 1998 TOK held a series of meetings with 'urban Māori' representatives to discuss a proposed development putea (fund). By February 1999 TOK concluded that there was sufficient support for the proposed optimum method for allocation to be reported to the Minister of Fisheries, and notified its intention to all interested parties. Immediately, opponents placed an injunction on TOK, preventing it from proceeding. In the mood of frustration felt by all, Tau Henare, Minister of Māori Affairs, told the May annual Māori commercial fisheries conference that ordinary Māori had 'had a gutsful' of bickering over allocation of fisheries assets and were sick of waiting for benefits from the settlement. He noted that the 1996 'deadline' had come and gone and TOK seemed to be no closer to an agreed system for allocation.[23] Then in August, at the annual meeting, TOK Chair Sir Tipene O'Regan claimed that acceptance by iwi of the latest allocation model was substantial. He said 63 percent of affiliated iwi now agreed, and that those trying to prevent allocation were a mix of some who wanted a different division, some who were opposed to the 'iwi principle', and some who were opposed to any allocation.[24] Sir Tipene also noted that the number of Māori employed in the fishing industry had risen dramatically. He was talking up the need for positive news because TOK had again failed to get agreement on the allocation model.

Tau Henare now called for nominations for new commissioners. In October 1998 Sir Robert Mahuta had been elected deputy Chair of the Commission. Sir Graham Latimer had retired both from TOK and as Chairman of Moana Pacific Fisheries Ltd in May 1998, and Phil Pryke had also resigned, so TOK was down to ten members. However, selection

OPPOSITE PAGE TOP

Area One Consortium's Dick Dargaville (left) and Treaty Tribes' spokesperson Tu Wylie appear happy after signing a TOK brokered agreement at the Hui-ā-Tau to work together to produce an equitable fisheries settlement allocation model July 1995.

TOK

OPPOSITE PAGE BOTTOM

Māori rights and environmental activist Mike Smith (centre with beret), Teanau Tuiono (left with beret) and others demonstrating outside the Wellington Events Centre (where the 26 July 1997 Hui-ā-Tau was being held) against the 1992 fisheries settlement. The protesters issued a press release pointing out that fighting by TOK commissioners over allocation of fisheries settlement assets over seven years (1989 to 1996) had cost more than $8 million. The protestors claimed that commissioners had been paid over $6 million in director and consultancy fees. The main point they claimed was that the fisheries settlement was wrong in principle in that it substituted the Treaty rights of all Māori for a corporate commercial structure. They were calling for the Commission to be sacked, for there to be no allocation of assets and for support for an independent process of reconciliation among Māori.

ATL: EP-1997-3055

of new commissioners was deferred until after the November 1999 New Zealand parliamentary elections.

The Court of Appeal then ruled that 'iwi', in terms of the Maori Fisheries Act, meant traditional tribes for the purposes of allocating assets. As well, the High Court ruled that commercial freshwater fisheries were included in the settlement. On 27 November 1999 the increasingly unpopular National Party government of Jenny Shipley lost the election to the Labour Party and its partners. Dover Samuels was appointed Minister of Māori Affairs, but resigned in June 2000. Parekura Horomia then became Minister of Māori Affairs. In August he appointed new commissioners from about 100 nominations. There were eleven commissioners in all. Four previous commissioners were retained (Naida Glavish, Craig Ellison, Shane Jones and Archie Taiaroa) and seven new commissioners were appointed (June Jackson, June Mariu, Ken Mason, Robert McLeod, Maui Solomon, Toro Waaka and Hon. Koro Wetere, Minister of Māori Affairs in the Lange Labour government). In addition, Shane Jones was elected Chair and Craig Ellison deputy Chair.[25]

Prior to becoming Chair, Shane Jones spoke with Prime Minister Helen Clark. She had taken a close interest in the fisheries settlement and was keen to see the matter of allocation settled as soon as possible after Labour had won the election. Her advice to him was to rise above the petty politics of the situation and keep his eye on the outcome: establishing an agreed method of allocating fisheries assets.[26] Thus began a new era for TOK. Could Shane Jones and the new Commission garner enough support for a resolution to the seemingly interminable squabbling over how fisheries assets should be allocated?

Like the previous year, this 1999–2000 year had been quiet and progress toward allocation of assets had been slow. However, progress was being made on other fronts. TOK was now able to recognise 78 iwi for the purposes of fisheries matters. The effort put into preparing iwi to carry out fisheries business was paying off. Financial performance was still considered excellent, and TOK's Charitable Trust continued to make grants for scholarships and other fishing related assistance.

Good news arrived in July 2001 when the Privy Council in London reaffirmed the decision of the Court of Appeal that allocation of pre-settlement assets must be to iwi, meaning traditional tribes. Robin Hapi, Chief Executive of TOK, welcomed this news, claiming that TOK had always been confident of its interpretation of the Maori Fisheries Act 1989 and the Treaty of Waitangi (Fisheries Claims) Settlement Act 1992, saying, 'It is time for Te Ohu Kai Moana to get on with allocating to representative iwi organisations the Pre-Settlement Assets and for representative iwi organisations to grow this economic base for the benefit of their members'. Hapi noted that the Privy Council had also been clear that litigation should stop, and reiterated the view of the Waitangi Tribunal that, 'Treaty matters are more for statesmen than lawyers'.[27] Following this milestone, the allocation of pre-settlement assets gained greater momentum. Challenges to TOK's 'Optimum Method of Allocation' could now be resolved, either through court action or through the dispute resolution processes. So in the latter part of 2001, negotiations between opposing parties continued in earnest.

In December 2001 TOK released *He Anga Mua – A Path Forward*, containing four separate allocation models, and called for submissions. Some twenty consultation hui were held throughout the country, with iwi representatives

OPPOSITE PAGE

Resource rentals imposed on fishers was one of the topics discussed at this Hui-ā-Tau held in the open at Ōrākei marae, Auckland, 30 July 1994. Speaking at the hui, TOK Chair Sir Tipene O'Regan said that a resource rental would destroy the potential benefits of the fisheries settlement and was contrary to Māori rights. The Crown decided not to proceed with rentals.

TOK

LEFT

Te Ohu Kaimoana chair Shane Jones and president of Nippon Suisan Kaisha (Nissui) shake hands after agreeing to establish a fisheries training initiative that will see two Māori each year selected to take up a 12-month programme in both NZ and Japan. Wellington TOK office, July 2001.

TOK: Tangaroa 62, 2001

BELOW

Francene Wineti working at NZ King Salmon Co. in Blenheim in 2004. Francene and Joseph Butterworth each won a prestigious Global Scholarship worth $250,000, funded by the partnership between the Treaty of Waitangi Fisheries Commission and Sealord's co-owner Nippon Suisan Kaisha Limited (Nissui). The scholarship supports two students for a year's intensive training in Tokyo.

TOK

seeking views on the options in the document. As well as the face-to-face meetings, TOK carried out an intensive publicity program in the media to try to hasten agreement. However, after an initial burst of activity from December 2001 to February 2002, certain groups went to court to prevent TOK consulting any further on these latest proposals. Their challenge was unsuccessful, but TOK took from the action that much more information should be made available to iwi, and provided this through further hui. Eventually, in August, TOK developed a preferred method of allocation for the whole fisheries settlement.[28]

After submissions were analysed in May 2002, TOK was ready to report the preferred solution to the Fisheries Minister as required. About the same time, in order to further hasten those reluctant iwi into an agreement, a group of 25 iwi released a report from the New Zealand Institute for Economic Research which claimed that Māori were losing over $1 million every month as a result of the delays. They called for the government to legislate to allow the immediate allocation of the assets.[29]

A further document, *Ahu Whakamua: Report for Agreement,* was prepared by TOK, outlining their preferred method of allocation that 'would see millions of dollars worth of fisheries assets allocated to iwi', as well as making Aotearoa Fisheries Limited (AFL) the largest fisheries company in New Zealand, and providing dividends and employment to iwi. Chair Shane Jones said that once iwi had met the structural and constitutional requirements laid out by TOK, they would receive their entitlement of fish quota and shares in AFL. The shares would generate annual income, as AFL would be required to allocate 40 percent of its annual profits to iwi.[30]

Te Ohu Kaimoana would be reshaped to replace the Treaty of Waitangi Fisheries Commission. It would oversee governance of AFL and the Te Putea Whakatupu Trust, which would hold a $20 million fund for the benefit of Māori who did not associate with an iwi. The commissioners of TOK would appoint the five directors of AFL and three trustees of the Trust. The functions of the new organisations would be similar to those of the current Commission, and include:

- Managing assets on behalf of iwi — either on a transitional basis, or until iwi wished to take on that management themselves after satisfying TOK criteria;
- Participation in commercial fisheries management;
- Assist with implementation of customary fisheries management;
- Foster research and development in fisheries; Develop fisheries policy for Māori;
- Receive and allocate any new quota as new species were introduced into the QMS.

On 14 August 2002 TOK convened a major hui at Hopuhopu, Waikato, at which it finally released its preferred method for allocation in *Ahu Whakamua: Report For Agreement*. The key to this allocation model was that both pre- and post-settlement assets were considered together. TOK noted that each category was estimated at about $350 million. There followed a further round of about twenty consultation hui at which 'in depth and robust debate' took place.[31] TOK was careful to meet also with those organisations that had taken court action.

Even so, Margaret Mutu claimed that 'bullyboy tactics' were used by TOK in trying to speed up the allocation of assets in late 2002. She wrote that:

The Commission ran a propaganda campaign which claimed that 91 percent of iwi representing 96 percent of Māori supported the Commission's allocation model. The mainstream media believed the propaganda, praising the chairman as the saviour of warring Māori factions. Māori lawyer and commentator Annette Sykes noted, 'Television, radio and newspaper stories that report the views of someone as highly placed as the Commission

chairman, Shane Jones, can combine to give those views far more authority and credence than they deserve . . . we've been provided with an illusion of informed consent, when the reality is something quite different.' Mainstream media subsequently ignored the fact that a large number of hapū and iwi took litigation to prevent the model being entrenched in legislation. The litigation failed and the matter was left in the hands of the Minister of Fisheries.[32]

Although there is a contrary view to Dr Mutu's statement, the fact is that years of bitter squabbling and expensive litigation came to an end in May 2003 when *He Kawai Amokura: A model for allocation of the Fisheries Settlement Assets* was presented to the Minister of Fisheries, Pete Hodgson, by the Chair of TOK, Shane Jones. According to TOK, the allocation report was supported by some 93 percent of iwi.[33] By August TOK was implementing the model by establishing Aotearoa Fisheries Ltd. 'The Minister of Fisheries confirmed that allocation proposals outlined in *He Kawai Amokura* will remain largely unchanged as it begins the Parliamentary process,' Chair Shane Jones stated.[34]

In November 2003 the Maori Fisheries Bill was introduced into Parliament. The legislation gave effect to the TOK's proposals for all inshore fisheries quota to be allocated to iwi through a coastline formula. Deep-water quota would be allocated to iwi through a formula based 75 percent on population and 25 percent on coastline. Shares in Moana Pacific Fisheries Ltd (owned by TOK) would be allocated on a population basis, and a substantial amount of cash would be distributed using a population formula, with each iwi guaranteed to receive at least $1 million. An electoral college would be established with nine members to choose the seven commissioners for the Board of TOK. In turn, these seven commissioners would appoint the directors of AFL.[35]

The Quota Management System surges

Surging ahead like a spring tide, on receiving the news of an agreement on allocation in August, the Fisheries Ministry introduced ten more commercial fish species into the QMS on 1 October 2002. The Minister of Fisheries, Pete Hodgson, announced the expansion, saying that there were now 54 species of fish under the QMS and the Ministry was working towards introducing a further 40 or more species over the next three years. Of course this was good news for Māori, who received 20 percent of the quota for each of these species. It also gave more 'certainty' to fishers with relevant catch history, who would now receive a tradable asset. So there was good positive news all around at the end of 2002.[36]

Earlier in the story it was noted that after allocation, the second important responsibility of TOK was to protect and grow the fisheries assets. The third of the key priorities was taking account of and giving effect to Māori customary fisheries, and getting Māori into the business of fishing was the fourth. These responsibilities are discussed below, going back to the original Treaty of Waitangi Fisheries Commission.

Growing the settlement assets

In 1994 TOK's strategy to protect and grow the assets they had won from the Crown was to further invest in the fishing industry through purchasing fishing quota and companies. In their 1995 annual report, it was noted that they had done

OPPOSITE PAGE

Lisa Rakuraku and Beryl Rogers with *Ahu Whakamua* documents explaining TOK's method of allocation of fisheries assets, Wellington headquarters, August 2002. Chair Shane Jones released a statement explaining the main features of the allocation method. Iwi were encouraged to read the documents and make their submissions.

TOK

OPPOSITE PAGE TOP

Shane Jones (right), Chair of TOK, delivers the Commission's report on an assets allocation model to Pete Hodgson, Minister of Fisheries, Wellington, May 2003.

ATL: Dominion Post *collection*

OPPOSITE PAGE BOTTOM

A Northland oyster farm, Whangaroa, June 1996. TOK purchased Pacific Marine Farms in 1996. This operation in Northland delivers export quality Pacific oysters to the world.

TOK

ABOVE

Former fish trawler *Newfoundland Breeze* is re-flagged and renamed *Taimānia*, Nelson, 1995. Daughter of TOK commissioner Shane Jones, Taimania, starred at the launch, where Sir Tipene O'Regan (Commission Chair) said *Taimānia* (meaning ocean horizons) symbolised freshness of youth, portending future promise for the fishing industry.

TOK: Tangaroa *27, 1995*

some analysis of aquaculture opportunities, had purchased a company with significant pāua quota (Salmond Smith Biolab), and were seriously considering further acquisitions.[37] To this end, highlights at the end of 1996 included an improvement in TOK's financial position (with assets of $554 million, up from $374 million in 1995), and further extension of the scholarships and research grants made available to get Māori into the business of fishing.[38] Similarly, in the 1996–97 year total assets again rose in value. Iwi were queuing up to undergo scrutiny in order to be ready to receive and utilise any assets to be distributed.[39] One of that year's important events was the Māori commercial fisheries conference, where TOK Chair Sir Tipene O'Regan hit out at critics who had claimed Māori had received no benefits from the fisheries settlement. Sir Tipene pointed out that over $60 million had already been distributed by way of training funds, quota leasing (at less than commercial rates to help iwi fishers get established) and involvement in customary fisheries issues.[40] For the 1998–99 year, Sir Tipene reported that the net value of assets had more than doubled (to near $1 billion) since the handover from the Crown some ten years ago.[41] The 1998–99 year was a good one for TOK, due to strong market conditions and high prices for fish products. The $15 million surplus that year made up for the net loss over the previous 1997–98 year. According to the TOK annual report, the net assets held by TOK rose to slightly over $400 million.[42] Further quota leases were issued at discounted rates to iwi and further training and development grants were issued to Māori to help them into the business of fishing.[43] The asset base continued to increase so that in 2002 it was $420 million.[44]

In 2000 TOK was faced with a major dilemma. Its partner in Sealord, Brierley Investments Ltd, wished to sell its shareholding in Sealord Group Ltd. TOK was keen to purchase those shares. Nissui (Nippon Suisan Kaisha Ltd) was finally selected to partner with TOK in the Sealord joint venture, with the promise of hundreds of new jobs and strategies for adding value along the fish-to-table supply chain.[45] The purchase by Nissui was completed in January 2001. According to Sir Tipene O'Regan, what is not so well known is that although the Brierley Joint Venture owned both shares and fish quota in Sealord, on transferring the half share to Nissui, TOK kept all the quota.[46]

The annual report for the 2000–2001 year declared a surplus after tax of $37 million and stated that the subsidiary companies (Sealord Group, Chatham Processing Ltd, Moana Pacific Fisheries Ltd, Prepared Foods Group and Pacific Marine farms Ltd) experienced 'strong trading'.[47] In a 'snapshot' of the New Zealand seafood industry, it was noted that the fish species hoki was the main export earner ($324 million), followed by greenshell mussels ($117 million) and rock lobster ($116 million). The main markets for these products were Japan, the USA and the European Union, all taking well over $200 million of products each.[48] Robin Hapi, Chief Executive Officer of TOK, noted that in the ten years since the 1992 fisheries settlement through to August 2002, the assets had built up to the extent that Māori owned or controlled, through the Commission, 33 percent of the New Zealand fishing industry.[49]

Securing Māori customary and traditional fishing rights

The customary and traditional fisheries joint Crown–Māori working party established in 1993, soon after the new Commission, had been 'slowly drafting regulations to reflect the nature and extent of Māori customary fisheries rights'. Agreed regulations were not expected until mid-1996. The TOK annual report noted with some cynicism that six years after the passing of the Maori Fisheries Act 1989, the Minister of Fisheries had announced the first two taiapure sited in Palliser Bay (South Wairarapa).[50] A taiapure is a local management 'tool' established in an area that has customarily

Mahinga kai, Te Roto o Wairewa (Lake Forsyth), 1948. While the men attend to the skinning and drying of eels, the women make flax baskets in which the eels will be transported when finally cured. Ngāi Tahu interests in traditional food and other natural resources and the places where those resources are obtained was of vital importance for survival after they had their land wrongly taken by the Crown.

ATL: F-040700. KV Bigwood

TOK customary fisheries delegation meets with Fisheries Minister Doug Kidd to report on 23 consultative hui they held throughout the country, Wellington, March 1994. From left: Hon. Ben Couch, Doug Kidd, Dr John Mitchell, Naida Pou and Shane Jones. Chair of the committee Archie Taiaroa is absent.

TOK

been of special significance to an iwi or hapū as a source of food or for spiritual or cultural reasons. Taiapure can be established over any area of estuarine or coastal waters.[51] The working party did, in July 1996, release a discussion paper outlining what they agreed were the nature and extent of the fishing rights guaranteed by the Treaty of Waitangi.[52]

There followed a long period of discussion and amendment and alleged obstruction by Crown members of the working party. In May 1997, Margaret Mutu had suggested that the regulations be submitted to the Waitangi Tribunal for their comment and that Māori only should be consulted and not, as Fisheries Minister Luxton wanted, other interested parties.[53] Nearly two years later, the new customary fisheries regulations came into effect on 1 February 1999, allowing, among other things, iwi to issue permits for customary seafood gathering.[54]

Preparing iwi for the business of fishing

TOK had decided early on that no assets would be given over to iwi before they were ready to receive them. Being ready meant having a representative 'body' with the appropriate staff and infrastructure and sufficient expertise to use the assets so allocated to the best advantage of their constituents. Tribes were defined as 'traditional' tribes, and not other Māori collectives such as urban authorities. Controversially, in February 1996 TOK released a full list of the iwi which would be recognised for fisheries purposes. An original list of 78 iwi to be so recognised had been decreased to 58 after TOK agreed to deal collectively with two groupings, fourteen Hauraki iwi and ten iwi of Te Arawa. The Arawa iwi were represented by Te Kotahitanga o te Arawa and the Hauraki iwi by the Hauraki Maori Trust Board.[55] TOK then provided funding and expertise to assist those to prepare themselves to participate in, and develop, the fishing industry. A total of some $3 million had been given out by 1996, when Te Ohu Kai Moana Charitable Trust was established. At the same time in 1996, TOK further refined its view of 'iwi readiness' and distributed a document, *Mandate Recognition of Iwi Organisations,* that set out minimum standards for an iwi organisation to be recognised.[56] It was now up to iwi to utilise the assistance available and ready themselves to take up the opportunities.

TOK had originally begun encouraging Māori into training and development for various tasks associated with fishing. Scholarships, grants and other support had been a feature from the very first years. In 2001 TOK announced plans to sponsor two Māori fisheries trainees jointly with Nissui in Japan. The training would include fisheries research, management, and technology. Funding for training was set aside in Te Ohu Kai Moana Charitable Trust, and several hundreds of thousands of dollars had been spent each year on these activities.

TOK's practice of annual tendering and leasing of quota to iwi as a means of assisting Māori into fishing was at times controversial. Some iwi took court action challenging some aspects of that practice. For example, in 2002 some iwi claimed that past tender leasing arrangements had been inequitable. But TOK declared that the courts had tested its methods and found them to be lawful. Other court action involved TOK's decisions to recognise certain groups as iwi and not others. There were also disputes over boundaries. Boundaries assumed more importance because they were partly used to calculate asset entitlement.[57] In May 1999 TOK again hosted a national Māori commercial fisheries conference, where it was reported that more iwi were being recognised 'for fisheries matters', and the total number had now reached 78.[58] In the 2001–2002 year TOK set aside $700,000 to assist those 78

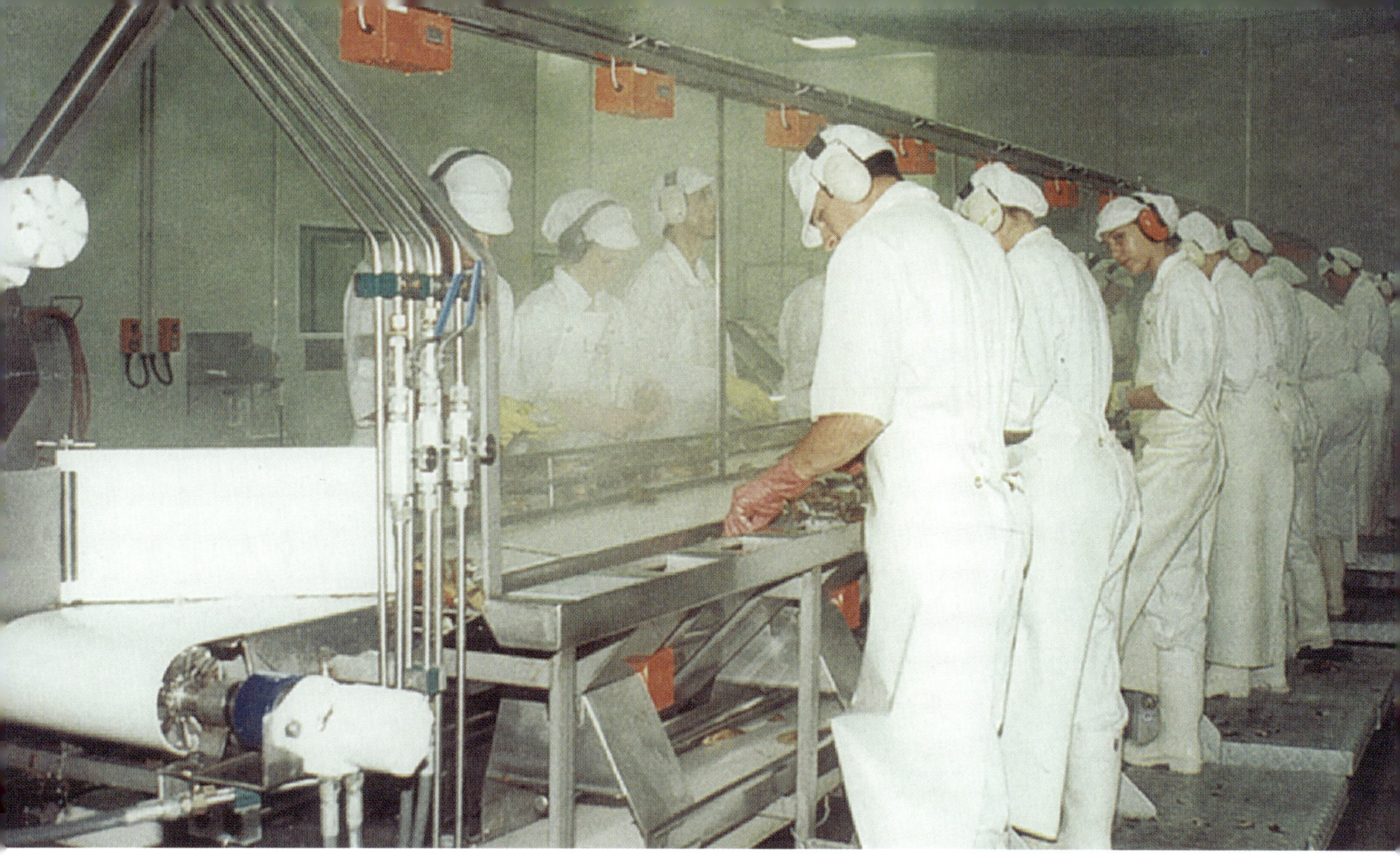

iwi organisations with involving, establishing and maintaining registers of beneficiaries, creating effective representative and communication methods, and bringing their organisations to a standard where fisheries assets could be effectively used and accounted for.[59]

Reading the sea of progress

In the tumultuous ten years from 1993 to 2003, TOK and its detractors, critics and opponents had spent vast amounts of time and huge sums of money to achieve 'equilibrium'. Even at the end of 2003, TOK still faced opposition to some of its policies and actions; but at least by then many tribes were either already fishing, or they were near to being able to 'handle' their share of the expected fisheries settlement assets. However, a Treaty history had already taught Māori that nothing is static, and that their rights had been under attack since the arrival of Pākehā. Once again the Crown was failing to address Māori rights in new circumstances. The focus shifted to the issues of coastal aquaculture and rights to the foreshore and seabed.

Meanwhile the whole New Zealand fishing industry had been expanding:

In the 1970s it was largely an inshore affair. Offshore waters, beyond our 12 nautical mile Territorial Sea, were fished by Japanese, Taiwanese, Korean, and Soviet vessels. With the introduction of the 200 nautical mile exclusive economic zone in 1977 and the introduction of the quota management system in 1986, many New Zealand companies went on to invest in fishing vessels to fish the available catch and onshore factories to process that catch. The industry had also grown from being a predominantly domestic supplier to one of the nation's leading export industries. In excess of 90 percent of all fish landed is exported. Today [2010] a small number of fishing companies provide the majority (about 80 percent of production) but there remain a large number of medium and smaller, usually inshore, fishing operations. In 2010 there were about 2200 individuals and companies owning quota and the value of that was around $3.5 billion, and there were in 2010 over 1500 commercial fishing vessels registered in New Zealand and 239 licensed fish receivers and processors.[60]

In 2003, Māori were just getting started.

OPPOSITE PAGE TOP

New $8 million Sealord shellfish processing factory in Nelson, March 1996. The factory employed some 150 people and exported greenshell mussels to about 20 countries.

TOK

OPPOSITE PAGE BOTTOM

Workers in a fish processing factory, Far North district, c. 1920. This contrasts with the modern Sealord shellfish factory in Nelson, opened in 1996, and capable of processing 15 thousand tonnes of shellfish a year.

ATL: 1/1-006318 Northwood brothers.

TOK's oyster farm, Pacific Marine Farms, in the Coromandel, 1997. It exports 500,000 dozen oysters to Australia, Taiwan, Hong Kong and Japan each year. Here, Henry Araiti pulls a stack of growing trays onto the harvesting barge.

TOK

ENDNOTES

1. David Johnson, *Hooked: The Story of the NZ Fishing Industry*. Hazard Press, 2004. p. 414.
2. Waitangi Tribunal, *Ngai Tahu Sea Fisheries Report*. Department of Justice, 1992. p. 245.
3. Waitangi Tribunal, 1992. p. 245.
4. TOK Annual Report, September 1993. p. 4.
5. Robin Hapi, former CEO of TOK, pers. comm., 2014.
6. TOK Annual Report, 1993. p. 9.
7. Ibid.
8. TOK, 'Pre-settlement assets: possible methods of allocation. A discussion document for Iwi'. 1993.
9. TOK Annual Report, 1994. p. 7.
10. http://www.manamoana.co.nz/Site/about/default.aspx
11. TOK Annual Report, 1994. pp. 5–6.
12. Tipene O'Regan, interview with Brian Bargh, 2015.
13. TOK Annual Report, 1994. p. 9.
14. Ibid, p. 10.
15. Ibid.
16. TOK Annual Report, 1994. p. 14.
17. *Tangaroa*, No. 26, August 1995. p. 1.
18. TOK Annual Report, 1995. p. 5.
19. Ibid, p. 7.
20. TOK Annual Report, 1996. p. 19.
21. TOK Annual Report, 1997. p. 19.
22. Ibid, pp. 9–10.
23. *Tangaroa*, No. 49, June 1999. p. 1.
24. *Tangaroa*, No. 50, August 1999. p. 1.
25. TOK Annual Report, 2000. pp. 4–5.
26. Shane Jones, pers. comm., March 2015.
27. Robin Hapi, CEO, TOK, Press Release, 3 July 2001.
28. TOK Annual Report, 2002. pp. 5–7.
29. Margaret Mutu, *The State of Māori Rights*. Huia Publishers, 2011. p. 95.

30. *Tangaroa,* No. 67, August 2002. p. 1.

31. TOK Annual Report, 2002. p. 7.

32. Mutu, 2011. pp. 128–29.

33. *Tangaroa,* No.73, October 2004. p. 3.

34. TOK, Press Release, 1 August 2003.

35. *Tangaroa,* No. 67, August 2002. p. 6.

36. Pete Hodgson, Press Release, 1 October 2002.

37. TOK Annual Report, 1995. p. 9

38. TOK Annual Report, 1996. p. 19.

39. TOK Annual Report, 1997. p. 19.

40. Ibid, p. 22.

41. *Tangaroa,* No. 49, June 1999. p. 5.

42. TOK Annual Report, 1999. p. 15.

43. Ibid, pp. 20–22.

44. TOK Annual Report, 2002. p. 14.

45. TOK Annual Report, 2000. pp. 10–11.

46. Tipene O'Regan, interview with Brian Bargh, 2015.

47. TOK Annual Report, 2001. pp. 7–11.

48. TOK Annual Report, 2000. p. 16.

49. *Tangaroa,* No. 67, August 2002, p. 2.

50. TOK Annual Report, 1995. pp. 14–15.

51. See sections 174–185 of the Fisheries Act 1996.

52. TOK Annual Report, 1996. p. 19.

53. John Luxton, Press Release, 28 May 1997.

54. TOK Annual Report, 1999. pp. 20–22.

55. TOK Annual Report, September 2003. p. 9.

56. TOK Annual Report, 1996. pp. 6–7.

57. TOK Annual Report, 1999. pp. 16–17.

58. TOK Annual Report, 1999. pp. 20–22.

59. TOK Annual Report, 2002. pp. 5–7.

60. Ministry for Primary Industries, http://www.fish.govt.nz/en-nz/Commercial/About+the+Fishing+Industry

SOMETHING NEW:
Who owns the seabed and foreshore?

The foreshore is defined as the intertidal zone, the land between the high- and low-water marks that is daily wet by the sea when the tide comes in. It does not refer to the beach above the high-water mark. The seabed is the land that extends from the low-water mark and on out to sea.[1]

In 1997 the Māori Land Court (Judge Heta Hingston) dropped a bombshell by ruling that customary Māori ownership of the foreshore and seabed in the Marlborough Sounds had not been subject to a 'blanket extinguishment'. This opened the way for iwi of Te Tau Ihu o te Waka (tribes of the top of the South Island) to try to prove that they were still owners of these areas and had rights to use and manage them. Dr John Mitchell, a Te Ohu Kaimoana (TOK) commissioner, said that their grievance went back to 1971, when the Marine Farming Act came into effect. That law had ignored traditional fishing grounds of the iwi and marine farming permits were issued over them. He said that the final straw came [in late July 1996] when [the Nelson-Marlborough local authority announced] a moratorium on new marine farming permits. This lasted until 1999.[2] Then in November, 2001 the Crown announced a national two-year moratorium on new permits, blocking iwi plans for marine farming. Moreover, the Crown declared that they would develop a tendering system for allocation of potential marine farming space, implying that the Crown owned the areas, which was disputed.[3]

The next year TOK supported iwi in Marlborough and Hauraki in their bids to have the Māori Land Court determine customary title to the foreshore and seabed in their areas.[4] In the Marlborough Sounds area there were a considerable number of marine farm permit applications after the lifting of the moratorium on granting them in 1999. Local iwi were concerned that these applications were over land they considered to be theirs by customary right. The Māori Land Court had ruled in October 1997 that it had jurisdiction to determine customary title over the foreshore and seabed and that Māori rights over that area had not been extinguished. The Crown appealed to the Māori Appellate Court, which in turn referred the case to the High Court. In June 2001, Justice Ellis in the High Court declared that the Māori Land Court did have jurisdiction to determine whether land down to the low water mark – the foreshore – was Māori customary land. He ruled, however, that New Zealand law did not recognise Māori customary title to the seabed. Te Tau Ihu o te Waka appealed the High Court's decision, and the hearing before the Court of Appeal was held in July 2002. The tribes were appealing the ruling that New Zealand law does recognise Māori customary title to the foreshore that borders dry land still owned by Māori, but once the title to adjacent dry land is determined, no further claim to the foreshore can be made. Te Tau Ihu o te Waka also disputed the ruling that the law does not recognise any Māori customary title to the seabed.[5]

Doug Kidd has said that he had a hand in the courts arriving at the position they did in respect of the seabed and foreshore. In his essay 'Parliament is Moving On',[6] he says that in 1992, when (as Minister of Māori Affairs) he was reviewing new Māori land legislation (Te Ture Whenua Maori Bill):

Foreshore and seabed hīkoi protest marchers enter Auckland on their way south to Parliament, April 2004.

Gil Hanly

[I] saw the opportunity, since this task of Minister of Māori Affairs had landed on my plate, to put things right about matters of customary title with respect to the foreshore and seabed. The great lawyers who had served before me in Morison Taylor and Co [Doug Kidd had worked there as a legal clerk, and the firm had produced Māori Land Court Chief Judge David Morison] were of the view that *Re The Ninety-Mile Beach* was wrongly decided in the New Zealand Court of Appeal. Much academic legal opinion was of the same view. And so I, in effect, repealed it in Te Ture Whenua Maori. I made the 'mistake' apparently of not spelling that out to each and every Member of Parliament. After the Ngati Apa decision in the Court of Appeal,[7] Prime Minister Helen Clark and her Deputy Michael Cullen claimed . . . 'Parliament didn't mean to do that.' But the record shows they all voted for it . . . Thus . . . my agreement to the drafting of the Māori Land Court jurisdiction to again allow the Court to investigate customary title to the foreshore and seabed led to the Ngati Apa Court of Appeal decision, which confirmed my intention to repeal the effect of *Re The Ninety-Mile Beach*. That led to Labour's enactment of the Foreshore and Seabed Act 2004.

The year 2002 also saw several tribes, with TOK support, challenge the Crown's proposed future management regime for marine farming. A claim was lodged with the Waitangi Tribunal that the government had failed to have regard to iwi interests in the coastal marine area and aquaculture. The hearing began in October 2002. TOK was also very concerned at government proposals to reform the law around marine reserves, as this appeared to impact negatively on the Māori fisheries settlement. So the Commission convened an iwi consultative group (Māori Fisheries Policy Forum) in September 2002 in order to monitor and advise on these reforms.[8]

TOK continued to support Māori in their battle with the Crown over the foreshore and seabed during 2003. In June the Court of Appeal ruled that the Māori Land Court did indeed have jurisdiction to determine whether or not the foreshore and seabed was Māori customary land.[9] In its ruling, the Court of Appeal found that 'native property rights are not to be extinguished by a side wind . . . The need for "clear and plain" extinguishment is well established and is not met in this case. In the *Ninety Mile Beach* case, the Court did not recognise that principle of interpretation'.[10] The ruling was foreshadowed by academic work done during the late 1980s and 1990s, which argued that the *Ninety Mile Beach* case was wrongly decided.[11]

The Labour-led government immediately announced its intention to pass legislation that would not allow the Māori Land Court to issue freehold title to foreshore and seabed areas and would ensure that such areas would be made 'public domain' – open to anyone. The debate was widely reported in the media. Public opinion was polarised with the National party opposition and other more extreme groups claiming that Māori would be able to stop the general public from going to enjoy the beaches.[12] TOK supported three national hui where those concerns were discussed.[13]

The Waitangi Tribunal held an urgent hearing and delivered its report in March 2004.[14] The Tribunal was critical of the government policy that would produce the proposed legislation:

> The policy clearly breaches the principles of the Treaty of Waitangi. But beyond the Treaty, the policy fails in terms of wider norms of domestic and international law that underpin good government in a modern, democratic state. These include the rule of law, and the principles of fairness and non-discrimination. The serious breaches give rise to serious prejudice:
>
> (a) The rule of law is a fundamental tenet of the citizenship guaranteed by article 3.
>
> Removing its protection from Maori only, cutting off their access to the courts and effectively expropriating their property rights, puts them in a class different from and inferior to all other citizens.
>
> (b) Shifting the burden of uncertainty about Maori property rights in the foreshore and seabed from

ABOVE

The main players arrive at the foreshore and seabed hui held at Ngā Whare Waatea marae, Mangere, 26 September 2003. From left: Mahara Okeroa (MP, Te Tai Tonga), Parekura Horomia (Māori Affairs Minister), Margaret Wilson, Trevor Mallard and John Tamihere. All the Māori Labour MPs except Tariana Turia supported the government despite widespread Māori disgust at their stance.

Fairfax NZ: John Selkirk MANGERE_ HUI_2_260903JS_32116 Dominion Post

OPPOSITE PAGE

Hīkoi marchers make their way across the Auckland Harbour Bridge, 27 April 2004. They were on their way to Wellington in protest against the proposed foreshore and seabed legislation.

Fairfax NZ: John Selkirk, BRIDGE_1_270404JS_79896 Dominion Post

the Crown to Maori, so that Maori are delivered for an unknown period to a position of complete uncertainty about where they stand, undermines their bargaining power and leaves them without recourse.

(c) In cutting off the path for Maori to obtain property rights in the foreshore and seabed, the policy takes away opportunity and mana, and in their place offers fewer and lesser rights. There is no guarantee to pay compensation for the rights lost.[15]

The Tribunal recommended to the government that their preferred option was for the Crown to 'go back to the drawing board, and engage with Māori in proper negotiations about the way forward'.[16] They added that if the government did proceed without changing the policy, Māori should be compensated for the loss of their rights in the foreshore and seabed.[17]

However, on 8 March 2004 Deputy Prime Minister Michael Cullen described the Waitangi Tribunal report on the foreshore and seabed as 'disappointing'. He rejected some of the central conclusions of the report: 'Those conclusions – particularly surrounding supposed breaches of the Treaty of Waitangi and the rule of law – depend upon dubious or incorrect assumptions by the Tribunal'. The most important of these was an implicit rejection of the principle of parliamentary sovereignty.[18]

During April and May 2004, a hīkoi (march of people) in support of Māori rights to the foreshore and seabed travelled from the Far North to Wellington, and was joined by thousands. Tariana Turia, a Minister in the government led by Helen Clark, spoke out against the proposed legislation and threatened to vote against it. She then resigned from the Labour government, leading soon after to the formation of a new political party: the Māori Party. Tariana Turia and Pita Sharples became its co-leaders. Other Māori Labour MPs were not so bold and chose to support their government's approach, much to the dismay and opprobrium of many Māori.

On 18 November 2004 the Labour-led government passed the Foreshore and Seabed Act, which declared that the land in question was owned by the Crown. Māori could, however, apply for 'guardianship' of certain areas. The Act was highly contentious, as an anonymous article in Wikipedia notes:

> The foreshore and seabed issue, as part of the larger race relations debate, was one of the most significant points of contention in New Zealand politics at the time, and remains a significant issue for many people in 2014. The Labour government's popularity was severely damaged by the affair, although subsequent polls showed that it recovered its support and Labour was elected for a third term in September 2005.[19]

While the Act was widely criticised by Māori, some iwi had chosen to negotiate agreements within the bounds of the Act. The first agreement made through the Act was ratified by Ngāti Porou and the Crown in October 2008.[20]

Māori anger and frustration with the Labour Government was made apparent when, in that 2005 general election, the new Māori Party won four Māori electorate seats from Labour.

A National-led government was elected to power in November 2008, with the Māori Party as part of the coalition. John Key became Prime Minister, and on 14 June 2010 he proposed the repeal of the Foreshore and Seabed Act. Repeal of the Act had been one of the Māori Party's conditions for partnering with National. The replacement Marine and Coastal Area (Takutai Moana) Act, proposed in late 2010, in turn created opposition from both sides of politics. Some Māori argued that this new legislation was a fraud, as essentially no Māori groups would meet the test for rights to the foreshore. Opponents, such as the Coastal Coalition, felt that the Bill put at risk free access to coastal areas for a large part of New Zealand.[21] The Coastal Coalition was formed to oppose the removal of coastal and marine land from Crown ownership and its coming under the 'control' of local Māori under the proposed legislation. It ran a number of media advertisements opposing the Bill.[22] The Māori Party supported the legislation and it was enacted.

The government releases its policy on the foreshore and seabed issue, Wellington, 17 December 2003. From left: Parekura Horomia, Prime Minister Helen Clark and Michael Cullen. The policy effectively ignored Māori protests and legal opinion that Māori owned these areas.

Fairfax NZ: Ross Giblin, Govt_5_1712_39861
Dominion Post

The Waitangi Tribunal meeting on the foreshore and seabed at the Westpac Stadium, Wellington, 20 January 2004. The Labour government ultimately rejected the Tribunal's findings in this case.

Fairfax NZ: Rob Kitchin, waitangi7_41930
Dominion Post

ABOVE

Protest on the last day of the Select Committee hearings on the Foreshore and Seabed Bill outside Bowen House, Wellington, 4 October 2004. Kate Lowe hands out leaflets.

Fairfax NZ: Phil Reid, castle_3_93483 Dominion Post

RIGHT

New Cabinet member Tariana Turia eyes her boss Helen Clark during the announcement of the Cabinet line-up, Wellington, 9 December 1999. Tariana resigned from the Labour Party in April 2004 over the Labour government's controversial foreshore and seabed legislation.

Fairfax NZ: Martin Hunter, 21-dpt-Jan2000-378 Dominion Post

ABOVE

Foreshore and seabed supporters. Right to left: long-time Māori rights activist Tame Iti (nearest camera), Rima Edwards (from Omanaia in South Hokianga, who passed away in April 2011), Sir Graham Latimer, Titewhai Harawira, and beyond her, Sanna Murray (from Te Hāpua), Parliament House, Wellington, 5 May 2004. They are at the steps of Parliament to show their disapproval of the Labour government's stance on denying Māori rights to the seabed and foreshore. The hīkoi of around 15,000 people arrived at Parliament after leaving Northland thirteen days earlier, picking up supporters on the way.

Gil Hanly

OPPOSITE PAGE

Pita Sharples (later to join Tariana Turia in the new Māori Party) leading protesters down Lambton Quay, Wellington, 5 May 2004. This was the protest march against the Labour government's foreshore and seabed legislation.

Gil Hanly

Aquaculture and marine farming

Before the 1990s, marine farming was a reasonably small industry in New Zealand, with small farms dotted around the coast, mostly in the Marlborough Sounds, Northland and around the Coromandel Peninsula. Over the next ten years, marine farming took off and demand for water space increased five-fold. The clean and nutrient-rich waters of New Zealand were rightly recognised as a great place to grow quality seafood quickly. By 2000, it had become clear that the existing legislation for planning and approving marine farms could not cope with this demand. There were two main complaints: marine farmers wanting new space were unhappy because of delays and costs in processing their applications; on the other hand, communities were concerned that the possible effects of marine farming were not being fully recognised and managed.[23]

During 2001 and 2002 Māori particularly those involved in aquaculture, and TOK became increasingly alarmed at the government's proposed aquaculture reforms. The Commission participated in the Marine Farming Claim to the Waitangi Tribunal on the basis that, in so doing, it was assisting Māori into the business and activity of fishing – a key component of its fundamental charter. The claim by Ngāti Kahungunu, Ngāti Whātua, Te Ātiawa ki te Tau Ihu, Ngāti Koata, Ngāi Tahu, and Ngāti Kuia pointed to conflict between the principles of the Treaty of Waitangi and the proposed aquaculture reforms.[24]

The Tribunal released its report on the claim, *Ahu Moana: The Aquaculture and Marine Farming Report*,[25] in December 2002. It found, among other things, that:

- Māori have a broad relationship with the coastal and marine area and therefore in aquaculture and marine farming;
- The proposed reforms would breach the principles of the Treaty;
- 'Māori have an interest in marine farming that forms part of the bundle of Māori rights in the coastal marine area';
- The Commission was the ideal body to facilitate a process of consultation and negotiation with the government to reach a fair settlement.

TOK carried out consultation and negotiations with the Crown over the report, but events were delayed due to the foreshore and seabed issue.[26] This did not stop the political opposition, led by the National Party, from being quick to criticise the outcome of those negotiations. Referring to the government's decision to give 20 percent of marine farming space to Māori, Phil Heatley, National's fisheries spokesperson, put out a press release asking, 'When will this madness end?' He added, 'There is no way you can interpret marine farming as a traditional Maori activity'.[27] Ironically, he became Fisheries Minister in the 2008 National Party-led government, and extolled the virtues of the eventual aquaculture settlement in a speech given at the agreement signing ceremony on 7 May 2009.[28]

The Labour-led government had felt that, without resolution, these Treaty claims would create uncertainty for the marine farming industry and local government decision-makers, because of the ongoing risk of legal challenge by Māori. So after the discussions with the Waitangi Tribunal claimants and TOK, it implemented the Māori Commercial Aquaculture Claims Settlement

OPPOSITE PAGE

Māori Party co-leader Tariana Turia and Prime Minister John Key announce their agreement to scrap the Foreshore and Seabed Act 2004, Parliament, Wellington, 14 June 2010.

Newspix.co.nz: 1080164 Mark Mitchell, NZ Herald.

Prime Minister John Key signing the 2009 Aquaculture Agreement on behalf of the Crown at Te Papa, on 7 May 2009. Phil Heatley (Minister of Fisheries), Bill English (Minister of Finance) and Chris Finlayson (Minister of Treaty Settlements) stand in order to his left. Nicholas Manukau (Ministry of Fisheries) looks on at far right.

Kirsty Woods

Act 2004 as a 'full and final settlement of Māori claims to commercial aquaculture on or after 21 September 1992'.[29]

The commercial aquaculture settlement provides iwi with assets equivalent to 20 percent of the water-space rights created in coastal waters since 21 September 1992. This includes the rights to 20 percent of any future new space allocated in Aquaculture Management Areas. Qualifying iwi get the rights to apply for a marine farming resource consent in a particular part of an Aquaculture Management Area. The possession of these rights gives iwi an opportunity to get into the business of marine farming. Iwi can sell their rights (subject to the agreement of 75 percent of iwi members voting at a properly constituted meeting), lease them, or use them to develop their own marine farming businesses, either by themselves or in association with other interests. Alternatively, iwi may choose not to do anything with these rights.[30]

At the time the aquaculture settlement was formalised in July 2005, the then Fisheries Minister David Benson-Pope praised the role that TOK had played in reaching a settlement, noting that it would play a new role in ensuring the benefits flowed to the relevant iwi. He believed that 'Aquaculture was deliberately left out of the 1992 Fisheries Settlement by the then National Government because it was "too hard" and it remained the unfinished business of that agreement'.[31]

Despite the best intentions of the Crown to implement its 2004 undertakings and obligations to Māori in the aquaculture sector of Coromandel and the South Island, it was not practical, given the existing aquaculture law and space provisions. Difficulties also arose in purchasing marine farming space for iwi. As a result, a formula was agreed that would provide a cash equivalent of the marine farming space. TOK was able to take the lead in developing a way to allocate the agreed cash compensation to benefit all eligible iwi. The settlement was signed at a ceremony at Te Papa Tongarewa (the National Museum of New Zealand) on 7 May 2009. Legislation giving effect to the agreement was passed through Parliament in March 2010. The settlement provided $97 million in full and final settlement of all Crown obligations for 'pre-commencement space' (aquaculture space) that was approved between September 1992 and December 2004, when Māori had a right to 20 percent of that space, but got nothing.

In 2011 the aquaculture settlement was amended to enable it to be delivered on a regional basis.[32] These regional agreements would in future be negotiated between the Crown and appropriate iwi organisations listed in the Maori Fisheries Act 2004. The Crown would transfer settlement assets to Te Ohu Kaimoana (the trustee), who will then allocate this to the appropriate iwi.[33]

In the end, Māori were served a partial victory. The aquaculture agreement with the Crown seemed fair and reasonable, and already TOK has seen some benefits. However, the foreshore and seabed issue has not been settled. Māori received a very poor and wholly unprincipled response from the Crown regarding the loss of their rights over those areas. Had the courts been allowed to rule on the various matters raised by Māori, there would probably have been a very different outcome.

NEXT PAGE TOP

Fred Te Miha, Dr John Mitchell and Robert McKewen at Te Ohu Kaimoana, signing the Iwi Allocation Agreement for the Marlborough and Tasman Regions on behalf of Ngāti Tama before signing the Aquaculture Agreement with the Crown on 7 May 2009. John Mitchell is a hero in the fisheries story for his tenacity in fighting for Māori rights to the seabed and foreshore – a fight yet to be settled.

Kirsty Woods

NEXT PAGE BOTTOM

Archie Taiaroa (left) and Ngahiwi Tomoana at Te Papa, signing the 2009 Aquaculture Agreement on behalf of Te Ohu Kaimoana.

Kirsty Woods

ENDNOTES

1. Waitangi Tribunal, *Report on the Crown's Foreshore and Seabed Policy*. Department of Justice, 2004. p. xi.

2. www.epa.govt.nz/AppendixD/Chronology in the Marlborough Sounds. Appendix D – Chronology of planning for marine farming.

3. *Tangaroa,* No. 41, February 1998. pp. 1–2.

4. Te Ohu Kaimoana (TOK) Annual Report, 1999. p. 19.

5. TOK Annual Report, 2002. pp. 14–15.

6. Maria Bargh (ed.), *Māori and Parliament: Diverse Strategies and Compromises.* Huia Publishers, 2010. pp. 189–97.

7. *Attorney-General v Ngati Apa* [2003] 3 NZLR 643.

8. TOK Annual Report, 2002. p. 16.

9. TOK Annual Report, 2003. pp. 23–24.

10. *Attorney-General v Ngati Apa* [2003] 3 NZLR 643 at 154.

11. See, in particular, Paul McHugh, 'Aboriginal Title in New Zealand Courts' 2 University of Canterbury Law Review, 1984. pp. 235–65; 'The Legal Status of Māori Fishing Rights in Tidal Water', 14 Victoria University of Wellington Law Review, 1984. pp. 247–273; Paul McHugh and Richard Boast, '*In Re Ninety Mile Beach* Revisited: The Native Land Court and the Foreshore in New Zealand Legal History', 23 Victoria University of Wellington Law Review 1993. p. 145.

12. http://en.wikipedia.org/wiki/New_Zealand_foreshore_and_seabed_controversy

13. TOK Annual Report, 2003. pp. 23–24.

14. Waitangi Tribunal, 2004.

15. Waitangi Tribunal, 2004. pp. xiv–xv.

16. Ibid.

17. Ibid, p. 143.

18. http://www.beehive.govt.nz/node/19091]

19. http://en.wikipedia.org/wiki/New_Zealand_foreshore_and_seabed

20. http://en.wikipedia.org/wiki/New_Zealand_foreshore_and_seabed

21. http://en.wikipedia.org/wiki/New_Zealand_foreshore_and_seabed

22. http://en.wikipedia.org/wiki/Coastal_Coalition

23. http://www.fish.govt.nz/en-nz/Aquaculture+Reform/Maori

24. http://www.fish.govt.nz/en-nz/Aquaculture+Reform/Maori

25. Waitangi Tribunal, *Ahu Moana: The Aquaculture and Marine Farming Report*, Department of Justice, 2002.

26. TOK Annual Report, 2003. p. 24.

27. National Party Press Release, 21 June 2004.

28. Phil Heatley, Press Release, 7 May 2009.
29. Ibid.
30. http://www.fish.govt.nz/en-nz/Aquaculture+Reform/Maori
31. David Benson-Pope, Press Release, 26 July 2005.
32. See Māori Commercial Aquaculture Claims Settlement Amendment Act 2011.
33. http://www.fish.govt.nz/en-nz/Aquaculture+Reform/Maori

THE NEW MILLENNIUM

The 2003 agreement on allocation and other issues by the Commission's affiliate iwi led to the Maori Fisheries Bill, which was considered by a Select Committee of Parliament in 2004.

Around 50 submissions were received, but the Committee made no significant changes.[1] On 16 September 2004 the Maori Fisheries Act was passed. The then Fisheries Minister, David Benson-Pope, said during the passage of the Bill that the legislation 'paves the way for economic growth that will benefit all New Zealanders'.[2]

This momentous event still had its sceptics and detractors. Ken Shirley, at the time an ACT Party MP, speaking during the passage of the Bill, said that 'patronising bureaucracies' would be set up under the legislation, and the model did not make commercial sense. However, Parekura Horomia, then Minister of Māori Affairs, countered the critics by declaring that the Act was 'a great moment for Māoridom'. He added, 'Paternalism is over. We don't need to be told how to run things'.[3]

On the day the Act was passed, the surviving architects of the 1992 Sealord deal gathered in the government caucus room to celebrate. Among those present were Sir Douglas Graham, Doug Kidd, Sir Tipene O'Regan, Sir Graham Latimer, Parekura Horomia, David Benson-Pope, Peter Douglas (Chief Executive Officer of Te Ohu Kaimoana, and former Māori advisor to the Prime Minister during the Sealord deal period), Shane Jones (Chair of the Commission), and former Prime Minister Jim Bolger. Jim Bolger said he considered the Sealord deal to be one of the crowning achievements of his administration. Shane Jones noted that the journey had been volatile, but now a new era had been ushered in. Sir Tipene O'Regan said that it was he who had proposed the idea to the then Minister of Māori Affairs, Doug Kidd, during a smoke break at a state function. Sir Tipene recalled that he told Doug Kidd he supported the Sealord deal possibility, but could promise to deliver only his iwi, Ngāi Tahu. It was Bob Mahuta (later Sir Robert Mahuta) who promised to get Māoridom behind the proposal. Sir Robert, who had passed away on 1 February 2001, and the other prominent Māori fisheries negotiator, Hon. Matiu Rata, were fondly remembered at the gathering.[4]

Ruth Berry, reporting on the parliamentary gathering for the *New Zealand Herald,* recalled a 1992 exchange between Matiu Rata and a kuia in Tainui that epitomised the clash within Māoridom over the signing of the deal – a clash which was as tempestuous as, and in many ways similar to, the foreshore debate. 'Matiu questioned whether she would prefer her kete of pipi over the economic wealth of the Sealord deal. In front of a large

NEXT PAGE TOP

Te Ohu Kaimoana meets, Bluff, 11 February 2002. From left: Chair Shane Jones, Sir Tipene O'Regan, and CEO Robin Hapi.

Fairfax NZ: Barry Harcourt, Fisheries_ commission_6089 Southland Times *6089*

NEXT PAGE BOTTOM

Donna Awatere-Huata, Doug Kidd (centre) and Ken Shirley listen to Harry Mikaere of the Treaty Tribes Coalition speaking on fisheries asset allocation, Wellington, 28 November 2001. It was another three years before allocation actually began.

Fairfax NZ: Anthony Phelps, 16-dpt-nov2001-954 Dominion Post

ABOVE

The first of two new Sealord freezer trawlers is renamed *Pākura* at a ceremony in Nelson, 1994. A second trawler *Aoraki* also joined the Sealord fleet within a month. Both were around 3000 tonnes.

TOK

LEFT

TOK lawyers Matanuku Mahuika and Damian Stone with some of the paperwork associated with current litigation brought by various Māori groups that threatened to delay progress on allocating fisheries settlement assets, TOK offices, Wellington, c. mid 1997.

TOK

hui, the kuia immediately responded: "E Matiu, mai ra ano e ora ana taku whanau i taku kete. My whanau have always got by with this kete of food".[5] A closer translation would be, 'since time immemorial my family have lived off the produce of this kete'.[6]

Following the passage of the new legislation in September 2004, the Minister of Māori Affairs appointed four new commissioners to what was then still the Treaty of Waitangi Fisheries Commission. Shane Jones was Chair and Craig Ellison, Ken Mason, Naida Glavish, June Jackson, Archie Taiaroa, Hon. Koro Wetere, Maui Solomon and Toro Waaka were the commissioners. They were joined by Harawira Gardiner, Rangimarie Parata, Dame Georgina Kirby and Rob McLeod.[7] In November 2004, the Commission was dissolved and all its undertakings were vested in Te Ohu Kaimoana (TOK) and Aotearoa Fisheries Limited (AFL). This final chapter of the story describes how Te Ohu Kaimoana set out to build on the work of past commissioners, staff and supporters to achieve its goals of getting Māori into the business of fishing, protecting and helping to grow the fisheries assets, and protecting Māori Treaty rights in all areas of fishing.

A new Chief Executive, Peter Douglas, had been appointed in March 2004. He had been advisor to Jim Bolger during the Sealord negotiations, and was therefore well informed of all the necessary changes after the Maori Fisheries Act 2004 became law. Peter declared that the Commission's work would be dedicated to proceeding with allocation. The first steps were to establish Aotearoa Fisheries Ltd, Te Ohu Kaimoana Trust and Te Ohu Kaimoana Trustee Ltd.[8]

In January 2005, as outlined in the previous chapter, the Maori Commercial Aquaculture Claims Settlement Act 2004 became law, creating the Takutai Trust. In March of that year, two new trusts – Te Wai Māori and Te Pūtea Whakatupu – were established.

The purpose of Te Wai Māori Trust is to advance Māori interests in freshwater fisheries through undertaking or funding research, development and education, and promoting the protection and enhancement of freshwater fisheries habitat. Te Pūtea Whakatupu Trust was established to promote Māori education, training and research. It achieves this by holding and managing the Trust Fund made available for these purposes. The Trust also exists to support and accelerate Māori social and economic development by providing strategic leadership in education, skills, and workforce development. The Takutai Trust is responsible for receiving aquaculture settlement assets from the Crown or regional councils, and allocating these assets to iwi via Iwi Aquaculture Organisations. The principal duties of the Takutai Trust are to allocate and transfer settlement assets; hold and administer settlement assets pending their allocation and transfer, and determine allocation entitlements.

OPPOSITE PAGE TOP

Twelve years on: The 1992 negotiators gather with Ministers Horomia and Benson-Pope to celebrate the passing of the Maori Fisheries Act 2004, Wellington, September 2004. From left: Hon. Mahara Okeroa, Sir Tipene O'Regan, Sir Graham Latimer, Hon. Doug Graham, Hon. Doug Kidd, former PM Jim Bolger, Commission Chairman Shane Jones, Hon. Parekura Horomia and CEO Peter Douglas.

TOK

OPPOSITE PAGE BOTTOM

Ngāti Ruanui representatives being congratulated by Te Ohu Kaimoana commissioners as the first iwi to have completed representation and structural requirements for allocation of Māori fisheries assets, TOK Wellington office, November 2002. From left: commissioners Craig Ellison and Archie Taiaroa, Ngāti Ruanui kuia Matekitawhiti Carr, commissioner June Mariu, Chair of Ngāti Ruanui Tahua Parata, commissioner Koro Wetere (handshaking), TOK CEO Robin Hapi, Ngāti Ruanui kaumātua Rocky Hudson, CEO of Ngāti Ruanui Spencer Carr, and Chair of Te Rūnanga o Ngāti Ruanui Pat Heremaia.

TOK

The New Millennium

In 2004 TOK took a 'snapshot' of the New Zealand fishing industry and the place of Māori in it. Some of the significant facts were that hoki was the most valuable commercial fishery: in 2002 it was worth about $300 million a year. Due to overfishing, the value of the 2003 annual catch dropped to about $200 million, and in 2004 the Minister of Fisheries used his powers to order a 45 percent cut in the Total Allowable Commercial Catch (TACC), in order for hoki fish stocks to rebuild. The hoki catch made up about one third of Sealord's business.[9] Later in 2004, kahawai and North Island eels were introduced into the QMS. These were traditionally a very important food source for Māori, and all New Zealand recreational fishers were concerned that commercial interests would dominate the kahawai fishery and reduce the fish that they valued for bait, sport and food. Quick to see the commercial potential, Aotearoa Fisheries Ltd bought a large commercial eel fishing company. There also were concerns among Māori that both eels and kahawai were still valued food sources, and that Māori customary rights could be compromised.[10] Māori received 20 percent of the quota of migratory fish, such as tuna, when they were brought into the QMS in 2004. At the same time, as recounted in the previous chapter, marine farming reforms were being promoted and Māori were arguing for their fair share of the marine farming space available.

Ngā Puhi fisheries asset settlement with the Commission is signed, Wellington, 22 September 2005. From left: Koro Wetere, CEO Peter Douglas, Chair Shane Jones, Teresa Tepania-Ashton and Ngā Puhi Chairman, Sonny Tau, signing the papers.

Fairfax NZ: Phil Reid, fish_3_127907 Dominion Post

Allocation proceeds at pace

To the great relief of everyone involved in Māori fisheries, in September 2005 the first six iwi were approved as mandated iwi organisations and the first Māori Fisheries Settlement assets were allocated to them. Ngā Puhi was the first iwi to be mandated (ready) to receive Te Ohu Kaimoana assets. Ngāti Kahu ki Whaingaroa (Northland), Te Aitanga a Māhaki (near Gisborne) and Ngāti Rārua (near Motueka) were also ready to be recognised by TOK and thereby receive their fisheries assets.[11] There were still disputes between iwi over boundaries, and until these were resolved TOK was not releasing assets, thus creating a big incentive to make compromises and be in a position to receive assets. Chief Executive Peter Douglas said that TOK would continue to offer assistance to iwi in reaching agreement with neighbours.[12] By March 2006 a further thirteen iwi had been mandated, making nineteen in total, and 27 percent of available assets had been allocated. By year's end 35 iwi had been mandated, with 60 percent of assets allocated.

The receipt of assets meant a great deal to all recipients. For example, Ngāti Porou received most of their allocation ($36 million in quota, shares in Aotearoa Fisheries Ltd and cash) in March 2006. Chair of the rūnanga, Apirana Mahuika, expressed his 'delight' and said, 'after all these years this is now the opportunity for us to move forward and to focus on reaping the benefits for Ngati Porou, as the runanga was at the forefront to advocate for and protect Ngati Porou fisheries interests and rights since 1989'. The rūnanga had established Ngāti Porou Fisheries Ltd in 2003.[13] Other iwi expressed similar sentiments once they had received their allocated assets. Progress on allocation had indeed been rapid. However, it was to be expected that some iwi organisations had been well established and well run over time and were in a position to easily satisfy the mandating criteria, while others were struggling and needed assistance.

In the period 2007 to 2009, 22 iwi were still to be mandated, but allocation activity was declining as the focus of TOK shifted to fisheries management. In August 2007 Shane Jones, then a Labour MP, resigned as a director of TOK. He had entered Parliament in 2005 as a list MP, and in October 2007 he was made a Minister by Prime Minister Helen Clark, making his dual role untenable. In November 2008 John Key became Prime Minister after the National-led coalition defeated the Labour-led government. Helen Clark resigned from Parliament and the relationship between government and TOK changed once again. New Ministers and MPs, some of whom had been critical of TOK and the Labour-led government's approach to Māori fisheries and rights, were now in the driving seat, and it was now twenty years since the first Māori fisheries legislation had been passed.

Expanding the assets

While allocation proceeded 'at pace', TOK was also responsible for oversight of the 'un-allocated' assets held in Aotearoa Fisheries Ltd (AFL). This company continued to acquire strategic assets, to expand, and to provide excellent returns to the shareholders. For example, in April 2007 AFL purchased a Whitianga-based fishing company (OPC) which held significant quota and other assets. The Chair of AFL at the time, Rob McLeod, said this acquisition further strengthened AFL's position in key North Island inshore fisheries.[14]

OPPOSITE PAGE

Apirana Mahuika (at podium) and Koro Dewes sharing a moment at the TOK annual conference, Pipitea marae, c. 2005. Seated, from left: Chair Shane Jones, Rob McLeod, Robin Hapi and Naida Glavish.

TOK

Looking back

In 2009, twenty years after the initial Māori Fisheries Settlement, the Chair of Te Ohu Kaimoana, Archie Taiaroa, was upbeat about the organisation and its achievements to that date. He observed:

> The change that is occurring in the Māori fisheries sector today is substantial. The challenges are exciting, and at times daunting. But they must be faced. It is a matter of some pride for me to be able to say that Te Ohu Kaimoana is addressing those challenges from the front foot. It is a mix of leadership and service; maintaining essential functions and vision for the future. I look back on the twenty years since the initial Māori Fisheries Settlement in 1989. The faces, the names, the stories. It is a living history of passions and rivalries; dreams and hopes for a future for our people, claiming their birthright as tangata whenua.
>
> In 2009, with the process of allocation and transfer of assets from the Māori Fisheries Settlement largely complete, it is good to step back and appreciate the work that has been done; the things that have been achieved.[15]

The 2010 Annual Report of TOK recorded that four iwi were yet to be allocated their settlement assets. The Chairman, Ngahiwi Tomoana, lamented this fact, describing it as 'unfortunate'. TOK noted, as it had done in most annual reports, that certain smaller fishing companies had been purchased and were being managed by Aotearoa Fisheries Ltd (the commercial arm of TOK).[16] As usual, the company reported a reasonable profit. It is interesting that TOK was facilitating quota management for those iwi who would rather lease their quota than actively fish it.

Sir Archie Taiaroa, who had first appointed to the Treaty of Waitangi Fisheries Commission after the signing of the Sealord deal in 2003, passed away on 21 September 2010. Another long-term fisheries commissioner, former Minister of Māori Affairs, Hon. Koro Wetere (first appointed to the Treaty of Waitangi Fisheries Commission in 2000) retired.[17]

By 2013, 55 of the 57 iwi (about 90 percent) had been mandated, and over $540 million of the Māori Commercial Fisheries Settlement had been transferred to these organisations.[18] Over this time, TOK had maintained a close relationship with all the iwi fishing organisations by disseminating information about important changes, threats and opportunities that had arisen since their establishment. With advice from these affiliates, TOK takes the lead in responding to the various issues that arise. For example, during 2013, a national fisheries management forum was established, designed to strengthen communications, address issues of common concern and research ways in which Māori could better participate in fisheries management.[19] Some of the issues dealt with by Te Ohu Kaimoana include: establishment of Marine Protected Areas; protection of endangered species (including dolphins, sea-lions, seabirds and sharks); stock assessment science; reviewing the total allowable catch of various species; and making various submissions to government. These tasks demonstrate the on-going complexity of the New Zealand fishing industry.[20]

OPPOSITE PAGE TOP

Te Atawhai Archie Taiaroa receiving his Order of Merit from Governor-General, Dame Silvia Cartwright (centre) for services to Māori at Ngapuwaiwaha marae (Taumaranui), 21 October 2009. Archie's wife Martha stands at left.

Wanganui Chronicle

OPPOSITE PAGE BOTTOM

Iwi Helpline staff Pania Tahau (at desk) and Megan Hall with Tony Sole, TOK manager of the exercise, prepare registration forms to send to Māori who claim to not know their iwi. Wellington headquarters, 1997.

TOK

In 2009, the issue of foreign charter vessels (FCVs) with foreign crews was brought into the public limelight when Sky Television released a TV documentary on the issue.[21] Foreign fishing vessels had been an industry-wide problem for years. Sealord and TOK, as major industry players, were open to be being percieved adversely. TOK was also aware that one of its main challenges was to get Māori into the business of fishing, yet charter boats were portrayed as having the opposite effect. It was alleged that foreign trawlers with crews mainly from Korea, Russia, and other countries [Indonesia] were chartered by New Zealand companies, to catch and process fish either on-board or in overseas factories, resulting in the loss of jobs here. It was claimed that the wage bill for 'the average New Zealand factory trawler' was some eight times that of a foreign trawler and that a large portion of our deep sea fish was caught by foreign boats and processed mainly in Thailand or China.[22] There was also evidence that foreign crew on some chartered boats were subject to appaling conditions akin to slavery and were often beaten and raped.[23]

A Ministerial Inquiry into the use and operation of Foreign Charter Vessels (FCV) was carried out during 2011, reporting to the Minister in March 2012. The inquiry found that:

> A small number of operators of foreign flagged FCVs have been mistreating their crews and acting in disregard of New Zealand's laws. These activities have put at risk New Zealand's standing in the international community and the reputation of the seafood industry . . . Although government has gone to considerable lengths in recent years to enforce New Zealand standards on board FCVs, it is clear that additional measures now need to be put in place to prevent the exploitation of foreign workers on FCVs, to safeguard New Zealand's international reputation and to protect the long-term interests of the fishing industry and 'Brand New Zealand'[24] . . .
>
> The use of FCVs is an important part of the fishing industry . . . provided the concerns that have been raised in this Report can be addressed, FCVs should continue to make a valuable contribution to the industry and New Zealand's economy.[25]

In January 2012, Sealord defended its use of foreign charters by taking media representatives aboard one of its charter vessels in Nelson to demonstrate its sea-worthiness and the conditions aboard. A spokesman for Sealord said, 'Chartering was a legitimate way to run a complex fishery in which vessels could be worth around $50 million'.[26] The Company also declared that observers would be sailing with its charter fleet in future to ensure standards are maintained.[27] In its submission to the inquiry, TOK said that FCVs mainly fished the low value, high volume species and were essential to utilising quota profitably.[28] They argued that, given the Government's compliance regime, FCVs operated in the same conditions as locally owned boats. 'If there has

OPPOSITE PAGE

Ramari Stewart stands beside a pan bone of a whale recovered at Moeraki, South Westland, 1994.

The Treaty of Waitangi Fisheries Commission hosted the 3rd Annual General Assembly of the World Council of Whalers in Nelson in November 2000. When Prime Minister Helen Clark became aware of the Commission's support, she forbade any government assistance. The Commission had attended previous conferences of the whalers and presented papers in 1998 and 1999. Papers from the Commission such as 'Maori Rights to Whale Use' and 'Beached Whales as Food' had angered conservationists here in New Zealand and overseas. Despite that, TOK staff and commissioners gave support to other peoples around the world who chose to continue to hunt whales, as was their customary practice. The Commission was clear that they did not support commercial whaling, but this message was lost in the uproar created by their support for traditional use.

TOK

been noncompliance (and we have no evidence to say that there is), then . . . The appropriate agencies should act to deal with the matter.'[29] This issue continues to be of concern.

This account of the Māori Fisheries story ends during the 2014–15 year. But the story itself has no ending. As long as there are fish to catch and mouths to feed, commercial fishing will continue to provide food for people of the world. Māori will continue to be a part of that story. The four pou, on which so much of this story is based – the Treaty of Waitangi, the New Zealand courts of law, the Waitangi Tribunal and Māori people, with their apparent almost infinite resilience – remain.

ENDNOTES

1. *Tangaroa,* No. 73, October 2004. p. 3.
2. *Tangaroa,* No. 73, October 2004. p. 1.
3. Ibid.
4. Ibid, p. 2.
5. *NZ Herald*, 17 September. 2004.
6. Robyn Bargh, pers. comm., 2015.
7. TOK Annual Report, 2004. p. 19.
8. *Tangaroa* No. 73, October 2004. p. 2.
9. Ibid, p. 4.
10. Ibid, pp. 12–13.
11. TOK Press Release, 23 September. 2005.
12. TOK Press Release, 5 October. 2005.
13. Ibid.
14. Aotearoa Fisheries Ltd Press Release, 2 April 2007.
15. TOK Annual Report, 2009. p. 4.
16. TOK Annual Report, 2010. p. 4.
17. Ibid, pp. 8–9.
18. TOK Annual Report, 2013. p. 9.
19. Ibid, p. 22.
20. Ibid, p. 11.
21. Sky Television, Press Release, 'The Great New Zealand Fishing Scandal', 14 July 2009.
22. Ibid.

23. http://www.stuff.co.nz/business/68739974/slavery

24. Ministry of Agriculture and Fisheries, 'Report of the Ministerial Inquiry into the use and operation of Foreign Charter Vessels', 2014. p. 31.

25. Ibid, p. 51.

26. http://www.stuff.co.nz/business/industries/6341357/Sealord, 2012

27. Ibid.

28. Te Ohu Kaimoana, 'Submission to the Review Panel on Foreign Charter Vessels', 7 October 2011. p. 9.

29. Ibid, p. 10.

AFTERWORD

Aotearoa New Zealand legends are rich in stories about the exploits of heroes, fish and the sea. The superhero Māui fished the stingray-shaped North Island up from the sea. The explorer Kupe discovered Aotearoa while chasing a huge octopus. Tangaroa, the revered god of the sea, was not to be disrespected. Even today, some fishers will return their first catch of a day to the sea, following the ritual practice of Māori in thanking Tangaroa.

These days New Zealand commercial fishers supply thousands of tonnes of fish to markets all around the world and fish can be purchased relatively cheaply in local markets. There are estimated to be over one million recreational fishers, all competing in some way for 'a feed of fish'. Māori now collectively control about half of the country's commercial fishing in one way or another. The fears and suspicions of the majority Pākehā culture over Māori ownership and control of fisheries, reflected in the actions of successive governments from those early years when competition for fish began, seem now to be largely laid to rest.

In 1869, only thirty years after the signing of the Treaty of Waitangi, two leaders from the Thames (Coromandel) area, Tanumeha Te Moananui and Aperahama Te Reiroa, with their supporters, petitioned Parliament. They were seeking protection of their fishing rights in the seas about their territories. The Thames sea area was where they and their forebears had lived, fished and thrived for several hundred years. But gold had recently been discovered in sands of the Thames estuary and in the rivers and streams flowing into it. Official documents show that the government was determined to allow gold prospectors and miners to have access to the area, which was still firmly occupied and owned by Māori.[1] The Crown insisted that it owned the foreshore and seabed. Māori, on the other hand, wanted to protect their foreshore and seabed for its rich shellfish and inshore fishery. Thus began a clash of values and laws between what the settler government believed was lawful and what Hauraki Māori believed was right. The petitions (translated for the official records) were couched in heartfelt language:

> O our messenger, on the ripple of the sea, to Wellington, to the Governor, to the General Assembly . . . The word has come that you are about taking our places from high water-mark outwards. The word has come that the Governor says he is to have these parts of the sea. O fathers, great is our grief, great is our sorrow, great is our objection . . . great is the considering of the heart on the subject of that work of yours. We have heard that you are a tribe of chiefs searching out good for this Island. That is not the work of chiefs, nor is it just work.

The statement explains that the petitioners gave the Crown 'the gold of Hauraki' and some land for a town (Thames), and allowed the Crown to supervise the leasing of the 'flats of the sea':

> O friends, it is wrong, it is evil. Our voice, the voice of Hauraki, has agreed that we shall retain the parts of the sea from the highwater mark outwards. These places were in our possession from time immemorial; these are the places from which food was obtained from the time of our ancestors . . . Leave to us our own, the places of

Afterword

the sea. Act justly towards the good tribe, because the searching for justice is with you . . . From all Ngatimaru, Ngatitamatera and Ngatiwhanaunga.

This petition was signed by Tan[a]meha Te Moananui.[2]

The Hauraki petitions did not invoke the Treaty of Waitangi, nor was it mentioned in the Parliamentary Committee report after the petitions had been considered. Rather, the government relied on the rights they believed they held under English common law (Crown prerogative over the foreshore). And so began a battle that escalated when, in 1870, the Crown began assuming control of fisheries. It would not end until 1992, when Māori knowingly, and somewhat willingly, gave up their rights to commercial fisheries in return for a major share of the New Zealand fishing industry and protection for their customary and traditional fisheries.

However, the battle for rights to the foreshore and seabed continues. Many Māori heroes and heroines have emerged, fighting to assert their fishing rights guaranteed by the Treaty of Waitangi.

In 1914 a woman, referred to by the Courts, newspapers and legal history as 'Waipapkura' or 'Wai Papakura', emerged as the heroine of the day. She was of Te Āti Awa, Taranaki. She was an 'activist' and is remembered for her resistance to the laws governing the catching of inanga, or whitebait, as they became known. Although her resistance was ultimately futile, her case[3] kept the struggle for Māori fishing rights alive, serving as an example to others who would follow.

When the legal breakthrough for Māori fishing rights came with the 1988 interim Māori Fisheries Agreement, opposition to this was muted. The Hon. Matiu Rata is reported to have said of the 1988 negotiations with the Crown that he and fellow Māori fisheries negotiators were near to 'one of the most historical decisions New Zealand is likely to make this century'.[4] Soon after, Chair of the new Māori Fisheries Commission, Tipene O'Regan, said of the 1988 interim settlement:

> It's not a final settlement but it's a chance to make a start and get structures into place ready for the rest of our 50 percent [of the New Zealand commercial fishery]. The return of our property rights in fish is one thing. Developing the skill to maintain and hold them for our mokopuna is the next great challenge.[5]

There was strong opposition from Māori to the 1992 Sealord deal, and it continued well after the Treaty of Waitangi Fisheries Commission had commenced work. Ten years later, when Māori had finally agreed on how to equitably allocate the assets of this pan-Māori settlement, opposition waned as Māori studiously accumulated more commercial fisheries assets. Their aim had been expressed many times since the Interim Agreement: to recover 50 percent of the commercial fisheries quota.

At the conclusion of the Sealord deal and resulting legislation, representatives of Māori and the Crown had a chance to review and reminisce. In ending his speech during the passage of the legislation through Parliament, Doug Graham referred to the final late night negotiation, saying, 'I think there was somebody else there, and I hope he remains with us'.[6] Sir Douglas Kidd said something

OPPOSITE PAGE TOP

Workers at the Hauraki Seafoods Ltd fish processing facility (owned by Hauraki Maori Trust Board) at Kopu near Thames. The Hauraki tribes see the factory as a modern-day extension of their commercial fishing activities carried on since settling the area.

TOK

OPPOSITE PAGE BOTTOM

Newly refitted 42.5m Sealord hoki trawler, *Thomas Harrison*, was launched by Fisheries Minister Doug Kidd, Nelson, late 1993. The trawler will provide fish for processing at the Sealord Nelson and Dunedin factories. It required 20 new staff working in two crews, each at sea for eight days.

TOK

Whina Cooper addresses the hui audience at Waitangi lower marae, Waitangi Day, 1985. Whina was at the forefront of Māori anger and protest from the 1960s through to the 1980s.

Gil Hanly

ABOVE

Sir Tipene O'Regan answers a question at the TOK Hui-ā-Tau at Ngāti Whātua o Ōrākei meeting house, 30 July 1994. His forthright debating style and vision for a Māori fisheries future were critical to the success of this story.

TOK

LEFT

Shane Jones and Prime Minister Jim Bolger, following the passing into law of the Maori Fisheries Act 2004, Wellington, 16 September 2004. Building on fifteen years of TOK work, extensive litigation and debate and millions of dollars in legal and other costs, Chair Shane Jones was finally able to get agreement from all the tribes to a method of allocation of fisheries settlement assets, which was then set in place by the Maori Fisheries Act.

TOK

Afterword

similar in 2014, when he recalled the brief and hectic negotiation period and the sudden decision of Carter Holt Harvey Ltd to sell out of the Sealord fishing company: 'It was just like something had come down from heaven, amazing timing'.[7]

During the debate in Parliament over the new Treaty of Waitangi (Fisheries Claims) Settlement Act 1992, which gave effect to the negotiated Sealord deal, Doug Graham said:

> I say that this nation owes a huge debt to those plaintiffs – if I may use that description [referring to the Māori Fisheries negotiators, Sir Graham Latimer, Bob Mahuta and Denese Henare, Hon. Matiu Rata and Tipene O'Regan]. I for one believe that in future annals of Māoridom, as the stories are told, those principal plaintiffs will shine forth as some of the great lights of their people.[8]

Rt Hon. Jim Bolger said in the same debate:

> I will never forget the meeting I held in the back room of the Prime Minister's office. I sat there with Sir Graham Latimer, Tipene O'Regan and Dr Bob Mahuta of Tainui. Those are the three who were at the first meeting with me . . . We sat down and asked, 'How do we make progress?' . . . and decided that we did have the opportunity and we could . . . capture the moment by assisting Maori with the purchase with their joint venture partners . . . of Sealord Products . . . There has been not a settlement like it anywhere in the world.[9]

The negotiators, Hon. Matiu Rata, Sir Graham Latimer, Sir Tipene O'Regan, Denese Henare and Sir Robert Mahuta, are today regarded by many as heroes, and a good number of the opponents of the Sealord deal are now closely involved in Māori fisheries.

OPPOSITE PAGE

Matiu Rata, previously MP for Northern Māori and Minister of Māori Affairs, continued the battle for Māori rights in fisheries until his death on 25 July 1997.

Fairfax NZ: fisheries-main-3

This book will not be the last word on Māori fisheries. On the horizon loom the dark clouds of global warming, which some predict will soon (in the next thirty years) irreparably alter, among other things, the acidity of the sea, the direction of ocean currents, and weather patterns at sea, leading to the demise of certain fish species and changing the migration patterns of fish around the world. As well, critics of New Zealand's Quota Management System say that the total allowable catch for many species is set far too high, and the availability of the most popular fish species has already declined to unacceptable levels. Anecdotal evidence from recreational fishers indicates that the fish being caught now are becoming steadily smaller and far less abundant than just a few years ago. For example:

> A spokesperson for recreational fishing group Legasea Hawke's Bay, Brian Firman, said a [boat launching] ramp survey of thousands of boaties over the past eight years had shown a 45 percent decline in the number of snapper caught and a 77 percent decline in groper catches. Mr Firman said over the same period Pania Surfcasting Club members reported a 70 percent decline in snapper catch and a 96 percent decline in gurnard from the shoreline. 'It has to be overfishing. We believe there's too much commercial pressure in the bay. Hawke Bay is a very flat and easily trawled area and over the years there has been too many fish coming out. There's also a lot of dumping and a lot of wastage with juvenile fish getting caught in the nets. A lot is not being reported and a blind eye is being turned to that and it needs to stop,' he said.[10]

Since Tanumeha Te Moananui and Aperahama Te Reiroa petitioned Parliament in 1869, Māori have been seeking acknowledgement and protection of their fishing rights. Now in 2015, this has been achieved to a considerable extent. However, demographic and environmental changes are putting huge pressure on the delicate fishing ecosystem. The challenge for Māori in the future is to conserve this ecosystem in order to ensure a viable and sustainable fishing industry.

ENDNOTES

1. Report of Committee On Thames Sea Beach Bill. *Appendices to the Journal of the House of Representatives* (AJHR) 1869 F-7, pp. 1–21.

2. *AJHR*, 1869 F-7, p 18.

3. *Waipapakura v Hempton* (1914) 33 NZLR 1065.

4. *Dominion,* 8 July 1988. p. 2.

5. *Tangaroa*, No. 1, September 1990. p. 1.

6. *Dominion,* 25 September 1992. p. 2.

7. Doug Kidd, interview with Brian Bargh, 2014.

8. Douglas Graham, *NZ Parliamentary Debates* (Hansard), V. 532, 3 December 1992, pp. 12842–843.

9. Jim Bolger, *NZ Parliamentary Debates* (Hansard), V. 532, 3 December 1992. p. 13036.

10. Peter Fowler, Hawkes Bay reporter, *Radio NZ News*, 15 December 2014.

APPENDIX

The Māori and English Texts of the Treaty of Waitangi

Māori Text

KO WIKITORIA te Kuini o Ingarani i tana mahara atawai ki nga Rangatira me nga Hapu o Nu Tirani i tana hiahia hoki kia tohungia ki a ratou o ratou rangatiratanga me to ratou wenua, a kia mau tonu hoki te Rongo ki a ratou me te Atanoho hoki kua wakaaro ia he mea tika kia tukua mai tetahi Rangatira – hei kai wakarite ki nga Tangata maori o Nu Tirani – kia wakaaetia e nga Rangatira Maori te Kawanatanga o te Kuini ki nga wahikatoa o te wenua nei me nga motu – na te mea hoki he tokomaha ke nga tangata o tona Iwi Kua noho ki tenei wenua, a e haere mai nei.

Na ko te Kuini e hiahia ana kia wakaritea te Kawanatanga kia kaua ai nga kino e puta mai ki te tangata Maori ki te Pakeha e noho ture kore ana.

Na kua pai te Kuini kia tukua a hau a Wiremu Hopihona he Kapitana i te Roiara Nawi hei Kawana mo nga wahi katoa o Nu Tirani e tukua aianei amua atu ki te Kuini, e mea atu ana ia ki nga Rangatira o te wakaminenga o nga hapu o Nu Tirani me era Rangatira atu enei ture ka korerotia nei.

Ko te tuatahi

Ko nga Rangatira o te wakaminenga me nga Rangatira katoa hoki ki hai i uru ki taua wakaminenga ka tuku rawa atu ki te Kuini o Ingarani ake tonu atu – te Kawanatanga katoa o o ratou wenua.

Ko te tuarua

Ko te Kuini o Ingarani ka wakarite ka wakaae ki nga Rangitira ki nga hapu – ki nga tangata katoa o Nu Tirani te tino rangatiratanga o o ratou wenua o ratou kainga me o ratou taonga katoa. Otiia ko nga Rangatira o te wakaminenga me nga Rangatira katoa atu ka tuku ki te Kuini te hokonga o era wahi wenua e pai ai te tangata nona te Wenua – ki te ritenga o te utu e wakaritea ai e ratou ko te kai hoko e meatia nei e te Kuini hei kai hoko mona.

Ko te tuatoru

Hei wakaritenga mai hoki tenei mo te wakaaetanga ki te Kawanatanga o te Kuini – Ka tiakina e te Kuini o Ingarani nga tangata maori katoa o Nu Tirani ka tukua ki a ratou nga tikanga katoa rite tahi ki ana mea ki nga tangata o Ingarani.

(signed) William Hobson, Consul and Lieutenant-Governor.

Na ko matou ko nga Rangatira o te Wakaminenga o nga hapu o Nu Tirani ka huihui nei ki Waitangi ko matou hoki ko nga Rangatira o Nu Tirani ka kite nei i te ritenga o enei kupu, ka tangohia ka wakaaetia katoatia e matou, koia ka tohungia ai o matou ingoa o matou tohu.

Ka meatia tenei ki Waitangi i te ono o nga ra o Pepueri i te tau kotahi mano, e waru rau e wa te kau o to tatou Ariki.

English Text

HER MAJESTY VICTORIA Queen of the United Kingdom of Great Britain and Ireland regarding with Her Royal Favor the Native Chiefs and Tribes of New Zealand and anxious to protect their just Rights and Property and to secure to them the enjoyment of Peace and Good Order has deemed it necessary in consequence of the great number of Her Majesty's Subjects who have already settled in New Zealand and the rapid extension of Emigration both from Europe and Australia which is still in progress to constitute and appoint a functionary properly authorised to treat with the Aborigines of New Zealand for the recognition of Her Majesty's Sovereign authority over the whole or any part of those islands — Her Majesty therefore being desirous to establish a settled form of Civil Government with a view to avert the evil consequences which must result from the absence of the necessary Laws and Institutions alike to the native population and to Her subjects has been graciously pleased to empower and to authorise me William Hobson a Captain in Her Majesty's Royal Navy Consul and Lieutenant-Governor of such parts of New Zealand as may be or hereafter shall be ceded to her Majesty to invite the confederated and independent Chiefs of New Zealand to concur in the following Articles and Conditions.

Article the first [Article 1]

The Chiefs of the Confederation of the United Tribes of New Zealand and the separate and independent Chiefs who have not become members of the Confederation cede to Her Majesty the Queen of England absolutely and without reservation all the rights and powers of Sovereignty which the said Confederation or Individual Chiefs respectively exercise or possess, or may be supposed to exercise or to possess over their respective Territories as the sole sovereigns thereof.

Article the second [Article 2]

Her Majesty the Queen of England confirms and guarantees to the Chiefs and Tribes of New Zealand and to the respective families and individuals thereof the full exclusive and undisturbed possession of their Lands and Estates Forests Fisheries and other properties which they may collectively or individually possess so long as it is their wish and desire to retain the same in their possession; but the Chiefs of the United Tribes and the individual Chiefs yield to Her Majesty the exclusive right of Preemption over such lands as the proprietors thereof may be disposed to alienate at such prices as may be agreed upon between the respective Proprietors and persons appointed by Her Majesty to treat with them in that behalf.

Article the third [Article 3]

In consideration thereof Her Majesty the Queen of England extends to the Natives of New Zealand Her royal protection and imparts to them all the Rights and Privileges of British Subjects.

(signed) William Hobson, Lieutenant-Governor.

Now therefore We the Chiefs of the Confederation of the United Tribes of New Zealand being assembled in Congress at Victoria in Waitangi and We the Separate and Independent Chiefs of New Zealand claiming authority over the Tribes and Territories which are specified after our respective names, having been made fully to understand the Provisions of the foregoing Treaty, accept and enter into the same in the full spirit and meaning thereof in witness of which we have attached our signatures or marks at the places and the dates respectively specified. Done at Waitangi this Sixth day of February in the year of Our Lord one thousand eight hundred and forty.

BIBLIOGRAPHY

Bargh, Maria. (ed.) *Māori and Parliament; Diverse Strategies and Compromises.* Huia Publishers, 2010.

Belich, James. *The New Zealand Wars and the Victorian Interpretation of Racial Conflict.* Auckland University Press, 1986.

Boast, Richard. 'Re Ninety Mile Beach Revisited: The Native Land Court and the Foreshore in New Zealand Legal History (1993)'. 23 Victoria University of Wellington Law Review, 1993.

Brown, R E. [Registrar-General]. 'Maori Population'. *Appendices to the Journals of the House of Representatives,* 1875, G-2.

Carkeek, Rikihana. *Home Little Maori Home, A Memoir of the Maori Contingent 1914–1916.* Totika Publications, 2003.

Caughey, Angela. *The Interpreter. The Biography of Richard 'Dicky' Barrett.* David Bateman, 1998.

Collingridge, Vanessa. *Captain Cook. The Life, Death and Legacy of History's Greatest Explorer.* Ebury Press, 2002.

Diamond, Paul. *A Fire in Your Belly: Māori Leaders Speak.* Huia Publishers, 2003.

Harris, Aroha. *Hīkoi: Forty years of Māori protest.* Huia Publishers, 2004.

Harris, Aroha. *Hīkoi: Der lange marsch der Māori* [Hīkoi: The long march of the Māori]. Orlanda Verlag, Berlin, 2012.

Harrison, Noel. *Graham Latimer – A Biography.* Huia Publishers, 2002.

Hill, Richard. *Maori and the State: Crown–Maori Relations in New Zealand/Aotearoa, 1950–2000.* Victoria University Press, 2009.

Johnson, David. *Hooked: The Story of the NZ Fishing Industry.* Hazard Press, 2004.

Keane, Basil. 'Kotahitanga'. In Maria Bargh (ed.), *Māori and Parliament: Diverse Strategies and Compromises.* Huia Publishers, 2010. pp. 9–15.

Keenan, E. 'A Maori Battalion: The Pioneer Battalion, Leisure and Identity 1914–1918'. Unpublished BA (Hons) thesis in history, Victoria University of Wellington, 2007.

Kidd, Hon. Sir Douglas. 'Parliament is Moving On'. In Maria Bargh (ed.), *Māori and Parliament; Diverse Strategies and Compromises.* Huia Publishers, 2010. pp. 189–197.

King, Michael. *Te Puea Herangi: From Darkness to Light.* Department of Education, 1984.

Lee, J. *The Old Land Claims in New Zealand.* Northland Historical Publications Society, 1993.

McHugh, Paul. 'Aboriginal title in New Zealand courts (1984)', 2 University of Canterbury Law Review, 1984.

McHugh, Paul. 'The legal status of Māori fishing rights in tidal water (1984)', 14 Victoria University of Wellington Law Review, 1984.

Mutu, Margaret. *The State of Māori Rights.* Huia Publishers, 2011.

New Zealand Law Commission. 'The Treaty of Waitangi and Māori Fisheries'. Prelim. Paper No. 9, 1989.

Orange, Claudia. *The Treaty of Waitangi.* Allen and Unwin/Port Nicholson Press, 1987.

O'Regan, Sir Tipene. 'Treaty Settlements, Fisheries and the Restoration of Rights'. Thomas Cawthron Memorial Lecture (unpublished), Nelson, August, 1999.

Parris, Robert. 'Memorandum to Under Secretary, Native Department'. *Appendices to the Journals of the House of Representatives*, 1881, G-3.

Pearse, Peter H. 'Fisheries Management Regimes for the Future'. XVIIIth Annual Conference of The European Association Of Fisheries Economists, Reykjavik, Iceland, 2007. http://www.univ-brest.fr/gdr- amure/eafe/eafe_conf/2007/peter_pearse_eafe2007.pdf

Pugsley, Christopher. *Te Hokowhitu a Tu: The Maori Pioneer Battalion in the First World War.* Random House, 2006.

Roa, Tom. 'Kingitanga'. In Malcolm Mulholland and Veronica Tawhai (eds.), *Weeping Waters: The Treaty and Constitutional Change*. Huia Publishers, 2010. pp. 165–174.

Salmond, Anne. *Two Worlds. First Meetings Between Maori and Europeans 1642–1772*. Viking, 1991.

Simpson, Tony. *Te Riri Pakeha. The White Man's Anger*. Alister Taylor, 1979.

Sinclair, Keith. *Kinds of Peace. Maori People After the Wars. 1870–85*. Auckland University Press, 1991.

Te Puni Kōkiri. *Tui, Tuia – Reflections on Māori Development 1984–2004*. Te Puni Kōkiri, 2004.

Te Ohu Kaimoana. Annual Reports, 2009–2014.

Treaty of Waitangi Fisheries Commission. Annual Reports, 1993–2006.

Waitangi Tribunal. *Manukau Report*. Department of Justice, 1983.

Waitangi Tribunal. *Motunui-Waitara Report*. Department of Justice, 1983.

Waitangi Tribunal. *Kaituna River Report*. Department of Justice, 1984.

Waitangi Tribunal, *Mangonui Sewerage Report*. Department of Justice, 1988.

Waitangi Tribunal. *Muriwhenua Fishing Report*. Department of Justice, 1988.

Waitangi Tribunal. *Ngai Tahu Sea Fisheries Report*. Department of Justice, 1992.

Waitangi Tribunal. *Fisheries Settlement Report*. Department of Justice, 1992.

Waitangi Tribunal. *Ahu Moana: The Aquaculture and Marine Farming Report*. Department of Justice, 2002.

Waitangi Tribunal. *Report on the Crown's Foreshore and Seabed Policy*. Department of Justice, 2004.

Walker, Ranginui. *Ka Whawhai Tonu Matou: Struggle Without End*. Penguin Books, 1990.

Ward, Alan. *A Show of Justice. 'Racial Amalgamation' in nineteenth century New Zealand*. Auckland University Press, 1995 (reprinted with corrections).

Ward, Alan. *An Unsettled History. Treaty Claims In New Zealand Today*. Bridget Williams Books, 1999.

Woon, R. W. 'Resident Magistrate's Report'. *Appendices to the Journals of the House of Representatives*, 1875, G-2, No. 13.

Websites

http://www.fish.govt.nz/en-nz/Aquaculture+Reform/Maori
http://www.fish.govt.nz/_indigenous_rights.pdf
http://www.fish.govt.nz/en-nz/Commercial/About+the+Fishing+Industry [Ministry for Primary Industries]
http://fs.fish.govt.nz
http://www.fourcorners.co.nz/new-zealand/new-zealand-history
http://www.manamoana.co.nz/Site/about/default.aspx
www.nzhistory.net.nz/people/matiu-rata
http://www.nzherald.co.nz/nz/news/article.cfm?c_id=1&objectid=170805
http://www.teara.govt.nz/en/waitangi-tribunal
http://en.wikipedia.org/wiki/New_Zealand_foreshore_and_seabed_controversy
http://en.wikipedia.org/wiki/Graham_Latimer
http://en.wikipedia.org/wiki/Doug_Graham
http://en.wikipedia.org/wiki/Coastal_Coalition
http://en.wikipedia.org/wiki/Wi_Parata_v_Bishop_of_Wellington

Court Cases

Wi Parata v the Bishop of Wellington (1877) 3 NZ Jur (NS) SC 72
New Zealand Maori Council and Latimer v Attorney General [1987] 1, NZLR 641
Ministry of Justice, High Court Judgement M662/85
Attorney-General v Ngati Apa [2003] 3 NZLR 643
Te Runanga o Wharekauri Rekohu Inc and ors v Attorney-General and ors (CA 297/92, judgment 3.11.92)
Waipapakura v Hempton (1914) 33 NZLR 1065

Parliamentary Records

New Zealand Parliamentary Debates (Hansard).
Appendices to the Journal of the House of Representatives (AJHR)

Bibliography

Acts of Parliament

1862	Native Land Act
1866	Oyster Fisheries Act
1877	Fish Protection Act
1884	Fisheries Conservation Act
1894	Sea Fisheries Act
1908	Fisheries Act
1962	Maori Community Development Act
1971	Marine Farming Act
1974	Maori Affairs Amendment Act
1975	Treaty of Waitangi Act
1983	Fisheries Act
1986	Fisheries Amendment Act
1989	Maori Fisheries Act
1992	Treaty of Waitangi (Fisheries Claims) Settlement Act
1993	Te Ture Whenua Maori Act
2004	Foreshore and Seabed Act
2004	Maori Commercial Aquaculture Claims Settlement Act
2004	Maori Fisheries Act
2011	Marine and Coastal Area (Takutai Moana) Act

Oral History Sources

Jones, Hon. Shane, interviewed by Brian Bargh in March 2015. Notes held by Te Ohu Kaimoana Trust.

Kidd, Sir Douglas Lorimer, interviewed by Brian Bargh in December 2014. Transcript and audio file held by Te Ohu Kaimoana Trust.

Latimer, Sir Graham Stanley, and Mahuta, Sir Robert Te Kotahi, interviewed by Adam Gifford in 1995. Transcripts and audio files held by journalist Adam Gifford.

O'Regan, Sir Tipene, interviewed by Brian Bargh in May 2015. Transcript and audio file held by Huia Publishers.

INDEX

Bold page numbers indicate photos, other illustrations and information in captions.

A

aboriginal title 14, 21, 23
allocation of benefits 127, 130–133, 135–145, **139**, **144**, 151–153, 179, 182, 184
Anderson, Judge 135
Anglican Church 13–14
Aotearoa Fisheries Ltd 75–76, 79–83, 96, 109, 130, 143–145, 181–182, 184
Aperehama, Manawa **64–65**
Aperehama, Rapine (Robin) **64–65**
aquaculture **136**, 148, 168–171, **170**, **172**, 179
Armitage, Russell 79
Ashlar Corporation 112
Awatere-Huata, Donna **176**

B

Baird, Fred 67
Ballard, Russ 86, 87
Bancorp 108–109
Banks, John 71
Baragwanath, David 66
Baragwanath, Owen 66
Barratt, Eric 120–123
Beatrix, Queen 87
Belgrave, Michael 88
Bennett, Sir John 73
Bennett, Bishop Manu 63, **64–65**
Benson-Pope, David 171, 175
Birch, Bill 106, 108
Bolger, Jim 71, 80, **92**, 99, 138, 175, 179, **195**
   in 1992 negotiations 86–88, 101, 105–112, 115–116, 120, **121**, 123, **178**, 197
Boyd, Wayne 109
Brierley Investments 86, 109–110, 112, 120, 124, 148
British colonial policy 13, 21–22
Brown, Bill **64–65**
Brown, Hone (Hone Rehu) **64–65**
Brown, Lewis (Nuk) **64–65**
Brown, Paihere 113
Brown, Raupo **64–65**

C

Canada 30
Carr, Matekitawhiti **178**
Carr, Spencer **178**
Carter, John 71
Carter Holt Harvey 86–87, 105, 110, 112, 120, 197
Cartwright, Dame Silvia **185**
Caygill, David 73, 80
Challinor, Rob 79, 109
Chatham Island Māori 115, 118
Chatham Processing Ltd 148
Chetwin, John 67
Clark, Helen 141, 159, 162, **163**, 165, 182
Coastal Coalition 162
commercial fishing industry 67–75, 87–88, 120–123, 127, 153
common law 21, 23–24, 55, 193
Cook, James 1, 5
Cooke, Sir Robin 93, 118–120, **119**
Cooper, Joseph **34**
Cooper, Whina 31, 32, **68**, 97, **100**, **194**
Couch, Ben Riwai 30–32, 124, 137, **150**

Court of Appeal decisions 11, 25, 48–49, 93, 118–120, **119**, 137, 141, 159
Cousteau, Jacques 63–66
Croft, Charlie 94
Cullen, Michael 159, 162, **163**
customary and traditional fisheries 13–16, 22–23, 49, 66, 109, 112, **149–150**, 157, 181
    commercial component 16–19, 43, 46, 49, 51
    recognition 60, 75–76, 115–116, 123, 127, 130, 135, 148–151
customary title 14, 46, 157–159

D

Dargaville, Dick 106, 112–113, 123–124, **139**
Delamere, Monita 63, **64–65**
Dewes, Koro **183**
Dewes, Whaimutu 79, 109, 124, 134
Douglas, Peter 105, 175, **178**, 179, **180–181**, 182
Douglas, Roger 63, **81**
Durie, Sir Eddie **29**, 30–32, 35, **38**, 39, 63, **64–65**, 70

E

Edwards, Katene 113
Edwards, Rima 113–114, **166**
eel fishing **18**, 35, **42**, 43, **47**, 50, 181
Elias, Sian 66
Ellis, Justice 135–137, 157
Ellison, Craig 79, 124, **128–129**, 141, **178**, 179
English, Bill **170**
Exclusive Economic Zone (EEZ) 130, 153

F

Federation of Commercial Fishermen 67
Fernyhough, John **81**
Finlayson, Chris **170**
Firman, Brian 197
Fish Protection Act 1877: 24, 46
Fisheries Acts 35–37, 48, 57–60, 86, 97
Fisheries Corporation 73
Fishing Industry Association 120–123, 127
Fishing Industry Board 67, 88
Fletcher Fishing 74, 79–80, 83, 112
foreign charter vessels 187–188
Foreshore and Seabed Act 2004: 159, 162

foreshore and seabed issue 157–162, **158**, **160–161**, **163–167**, 191–193
freedom of the seas principle 24

G

Gardiner, Wira 96, 106, 116, 179
Gillanders-Scott, Kenneth 30
Glavish, Naida **128–129**, 141, 179, **183**; see also Pou, Naida
Goldsmith, Quentin **136**
Goodall, Maarire **64–65**
Graham, Doug 80, **92**, 99, **100**, 175
    in 1992 negotiations 88, 101, 105–106, 110, 115–116, **117**, 123, **178**, 193, 197
Gregory, Bruce 73
Greig, Justice 66–67

H

Habib, George **107**
Hadfield, Octavius 14
Hall, Donna **136**
Hall, Megan **185**
Hanan, Ralph 27, 30
Hancock, Bruce 109–110
Hape, Tiopira Te Rauna **132–133**
Hapi, Robin 80, 109, **128–129**, 141, 148, **176**, **178**, **183**
Harawira, Titewai 27, **166**
Harman, Paul **64–65**
Harris, Aroha 32
Hauraki Māori 133–134, 151, **192**
Heatley, Phil 168
Henare, Denese 67, 71, 88, 105, 197
Henare, Sir James 73
Henare, Tau 137–138
Heremaia, Pat **100**, **178**
Heremaia, Steve **100**
Heron, Justice 118
Higgins, David 112, **132–133**, 134
Hingston, Judge Heta 94, 157
Hobson, William 1
Hodgson, Pete 145, **146**
Hohepa, Patu 27, 30
Horomia, Parekura 141, **161**, **163**, 175, **178**
Hudson, Rocky **178**

Index

I
Ihaka, Roera 113
Ihaka, Tangi **64–65**
indigenous peoples 11
Iti, Tame **166**
iwi as asset recipients 75, 79–83, 118, 130–133, 135–138, 141–153, 171, **172**, **178**, 179, 182–184

J
Jackson, June **128–129**, 141, 179
Jackson, Moana 118
Jarman, Nick 79, 124
Jennings, Stephen 79
Jones, Shane 39, 63, 66, 124, **128–129**, **150**, 182
 in 1992 negotiations 105, 112–114, 116
 as TOK chair 141–145, **142**, **146**, 175, **176**, **178**, 179, **180–181**, **183**, **195**

K
Kaituna River inquiry and report 13–14, 35
Kapa, Arthur 113
Kapa, Geranium 113
Karaka, Arama 48
Karetai, William 60
Kawiti, Eddie **68**
Kenderdine, Shonagh 63, 66
Key, John 162, **169–170**, 182
Kidd, Doug 55–56, 59, 80, **98**, 120, 127, **150**, 175, **176**
 in 1992 negotiations 85–88, 101, 105–112, 115–116, **117**, 123–124, **178**, 193–197
 profile 97–99
 view on foreshore and seabed rights 157–159
Kirby, Dame Georgina 179
Kirk, Norman 27–32, **28**
Kororareka 6

L
Labour Government 1972–75: 27–35
Labour Government 1984–90: 25, 37–39, 58, 67–75, 80, 93, 116–118
Labour-led Governments 1999–2008: 141, 157–162, **163**, 168–171, 175, 182
Labour Party 10, 89, 115
land losses 6–10, 14–15, 19, 25–27, 30, 32–35, 43, 46, 63, 91
Land March, 1975: **7–8**, 32–35, **33–34**, **100**
Lange, David 25, 32, 37, 58, 71, **72**, 80, 112
Latimer, Sir Graham 63, 67, 71, 73, 79–83, **81**, 124, 138, **166**
 in 1992 negotiations 86–89, 96, 101, 105–113, **117**, 175, **178**, 197
 as Māori Council leader 25, 27, 66–67, **95**
 profile 94–96
 as Waitangi Tribunal member 30, 35, **38**, 63, **70**
Law Commission 21–25, 48, 58
Legasea Hawke's Bay 197
Local Government Commission 30
Love, Sir Ralph 73
Luxton, John 130, 135, 151

M
Magna Carta 23–24
Mahuika, Apirana 182, **183**
Mahuika, Matanuku **177**
Mahuta, Sir Robert (Bob) 67, 91–93, **92**, **107**, 124, 138, 175
 in 1992 negotiations 87–88, 101, 105–112, 116, 197
Major, Phil 106
Mallard, Trevor **161**
Mana Motuhake 72
Mangonui Sewerage Scheme inquiry and report 39, 63
Maniapoto 137
Manu Motuhake 89
Manuera, Tuhoe **119**
Manukau inquiry and report 13–14, 19–20, 35–37
Maori Affairs Act 1967 and amendments 30, 89
Maori Commercial Aquaculture Claims Settlement Act 2004: 168–171, 179
Māori Congress 123–124
Māori Council 25, 27, 49, 60, 66–67, 88, 94–96
Māori Development Corporation 83
Maori Fisheries Act 1989: 74–76, 141
Maori Fisheries Act 1996: 130
Maori Fisheries Act 2004: 145, 171, 175, **178**, 179

Māori Fisheries Commission 75–76, 79–83, 87–88, 110, 120, 124, 127
Māori Land Boards 30
Māori Land Court 157
Māori parliamentarians 41, 46–48, 51, 55, 71, 123, **161**, 162
Māori Party 162
Māori Women's Welfare League **31**, 32
Marine and Coastal Area (Takutai Moana) Act 2010: 162
marine farming 157–159, 168–171, 181
Marine Farming Act 1971: 157
Mariu, June **128–129**, 141, **178**
Marsden, Maori 63, **68**
Martin, Bob 67, 72–73
Maruera, Maimau **100**
Mason, Ken 141, 179
Matiu, Makari 113
Matiu, McCully 63
McDonald, Ken 79
McIntyre, Duncan 27, 59
McKewen, Robert **172**
McKinnon, Don 116, **117**
McLeod, Robert **128–129**, 141, 179, 182, **183**
McMath, Witi 113, **132–133**
McMullin, Justice **119**
Meurant, Ross 110
Mikaere, Harry 66, 79, **132–133**, 134
Mitchell, John 79, 124, **150**, 157, **172**
Moana Pacific Ltd **82**, 83, 130, 138, 145, 148
Moore, Mike 80
Morgan, Paul **132–133**
Motunui-Waitara report 14, 35
Moyle, Colin 59, 67, 71, 73–74
Muaupoko 118
Muldoon, Robert 32, 55, 58
Muriwhenua 105, 113–114, 137
Muriwhenua inquiry and report 13–14, 16–20, 23, 26, 39, 41, 48–51, 57, 60, 63–71
Murray, Errol 113
Murray, Sanna **166**
Mutu, Margaret 118, **119**, 135, 143–145, 151

N
Nathan, E.D. (Ned) 63
National Governments pre-1975: 27
National Governments 1975–84: 30, 55–58
National Governments 1990–99: 80, 85–88, 93, 97–101, 105–112, 115–124, 138, 141, 171
National-led Governments 2008–2014: 162, 168, **169–170**, 182
National Party 35, 110, 116, 123, 159, 168
Native Land Court 6
Native Lands Act 1909: 9
Nepia, Billy 60
Ngā Puhi 80, 113–114, 137, **180–181**, 182
Nga Tamatoa 30
Ngāi Tahu 60, 66–67, 72, 75, 80, 83, **92**, 94, 99, 105, 108–112, 115, **149**, 168
Te Rūnanga o Ngāi Tahu 134
Ngāi Tahu Fisheries Ltd 83
Ngāi Tahu Sea Fisheries inquiry and report 13–20, 22–23, 41–51, 75, 85, 88
Ngāi Tamanuhiri Whānui Trust 133
Ngāiterangi 137
Ngata, Sir Apirana 27
Ngata, Sir Henare 73
Ngatata, Wiremu Tako 48
Ngāti Apa 115, 159
Ngāti Awa 118
Ngāti Kahu 63, **119**
Ngāti Kahu ki Whaingaroa 182
Ngāti Kahungunu 115, 118, 133–134, 168
Ngāti Koata 168
Ngāti Kuia 168
Ngāti Pikiao 14
Ngāti Porou 118, 162, 182
Ngāti Rārua 182
Ngāti Ruanui **100**, **178**
Ngāti Tama **172**
Ngāti Toa 14, 115
Ngāti Wai 32, 113
Ngāti Whātua **68**, 110, 115, 118, 168
Ninety Mile Beach 113, 159
Nippon Suisan Kaisha (Nissui) **142**, 148, 151
Norman, Waerete **64–65**

Index

Normanby, Lord 13
NZ King Salmon **142**

O

Okeroa, Mahara **161**, **178**
Orakei Parliaments 9, 48, 49
Ord Minnett 112
O'Regan, Roland 93–94
O'Regan, Sir Tipene 25, 59–60, 67, 71–74, 79–80, **81**, **92**, 99, 134–135, **176**, **178**
   profile 93–94
   and Sealord deal 85–88, 101, 105–116, **111**, **117**, 123, 175, 197
   as TOK chair 124, 137–138, 148, 193, **195**
overfishing 49, 56–57, 197
Oyster Fisheries Act 1866: 4, 17–19, 41
oyster gathering and farming **15**, 17–19, 37, 46, 51, **146**

P

Pacific Marine Farms **146**, 148
Paku, Tom **132–133**
Palmer, Geoffrey 48, 71, **72**, 73, 80, 89
Paranihi, Lisa **132–133**
Parata, Rangimarie 179
Parata, Tahua **178**
Parata, Tame 41
Parata, Wiremu Te Kakakura 13–14
Paul, Maanu 71, 74, 112–113, **117**
Pearse, Peter 56
Peters, Wayne **132–133**
Peters, Winston 67, 71, 80, 85, 123
Pou, Naida 124, **150**; see also Glavish, Naida
Prebble, Richard 73–74, **81**
Prendergast, James 13–14
Prepared Foods Group 148
Privy Council 11, 48, 137, 141
protests by Māori 5, 6–11, 32–35, **33–34**, **36**, 41, 48–51
Pryke, Philip 124, 138

Q

Quota Management System 25, 49, 55–60, 66–67, 71, 73, 75, 93–94, 120–123, 145, 153, 181, 197

Māori quotas 73–75, 79–83, 85–88, 110, 123, 127, 135, 145, 148, 151, 181

R

racism 55–56
Rakuraku, Lisa **144**
rangatiratanga 1, 13, 17, 22, 37, 51, 66, 135
Rangiheuea, Anaru 124
Rangitāne 115, 118
Rankin, Graham 113–114
Rapata, Rapata (Kut) **64–65**
Rata, Matiu 27–35, **28**, **36**, **64–65**, 66–67, **69**, 71–74, **90**, 138, 175–179, 193, **196**
   in 1992 negotiations 86–88, 101, 105–112, **107**, 115, **121**, 197
   profile 89–91
Ratana Movement 10, 89
Reedy, Amster **69**, 112
Reedy, Tamati 123–124
Rei, Matiu 115
Renata, Toko 134
Repudiation Movement 6
reservation of fishing grounds 19, 22, 46, 49, 75, 86, 123
Reweti, Brownie 30
rock lobster fisheries 14, 75, **82**, 83
Rogers, Beryl **144**
Rotorua 35
Rowling, Bill 32, **34**, **100**
Royal Greenland 87, 109, 112
Ruruku, Pene 108
Ryder, Eugene **136**

S

Salmond, J.W. 48
Salmond Smith Biolab 148
Samuels, Dover 113, 141
Sayers, Ross 67
Scott, John 134
Scott, Ken 30
sealers and whalers 5

Sealord Products/Sealord Group 80, 83, 86–88, 97, 105–124, **111**, **121**, 127, 148, 193
    operations and vessels **131**, **152**, **177**, 181, 187–188, **192**
Sewell, Henry 25
Sharples, Pita 162, **167**
shellfish gathering 4, 16, 20, 35, **44–45**
Shipley, Jenny 138–141
Shirley, Ken 175, **176**
Skeggs 83
Smith, Mike **139**
Sole, Tony **185**
Solomon, Bill 94
Solomon, Maui 118, **128–129**, 135, 141, 179
Sorrenson, Keith 63
Southwich, Laurie 30
Stannard, Bob 88
state-owned asset sales 63, 80, **81**, 93
Stewart, Ramari **186**
Stone, Damian **177**
Stout, Sir Robert 48
Sykes, Annette 79, 99, 143–145
Szaszy, Mira **31**, 63, 79, 113, 124

T

Tahau, Pania **185**
taiapure 75, 86, 148–151
Taiaroa, Archie 124, **128–129**, 141, **172**, **178**, 179, 184, **185**
Taiaroa, H. K. 41
Tainui 71–72, 88, 91–93, 99, 112, 113, 116
Tainui Māori Trust Board 67, 71–72
Taituha, Pumi 116
Takitimu District Māori Council 71
Tako, Wi 48
Takutai Trust 179
Talley, Peter 71, 88
Tamihere, John **161**
Taonui, Aperahama 114
Tapsell, Peter 86, 88, 123
Tasman, Abel 1, 5, 87
Tau, Henare 67
Tau, Rakiihia 135

Tau, Sonny **180–181**
Taumata Paepae 137
Tawhiao, King 6, 9
Taylor, Aila 35
Te Aitanga a Māhaki 182
Te Arawa 151
Te Atairangikaahu, Dame **92**, 116
Te Āti Awa 14, 35, 193
Te Ātiawa ki te Tau Ihu 168
Te Heuheu, Georgina 63, **100**
Te Kotahitanga Parliament 9
Te Moananui, Tanumeha 191–193, 197
Te Ohu Kaimoana (TOK) Treaty of Waitangi Fisheries Commission 124, 127–154, **128–129**, 175–188, **176–177**, **183**, **185**, 193
    in foreshore and seabed controversy 157–159
    marine farming issues 168–171
Te Ohu Kaimoana Charitable Trust 151
Te Ohu Kaimoana Trust 179
Te Ohu Kaimoana Trustee Ltd 179
Te Puea, Princess 91, 93
Te Pūtea Whakatupu Trust 143, 179
Te Reiroa, Aperahama 191, 197
Te Roopu o te Matakite 32
Te Tau Ihu o te Waka 157
Te Ture Whenua Māori Act 1993: 99, 157–159
Te Wai Māori Trust 179
Te Weehi, Tom 23, 48–49, 60
Te Whānau ā Apanui 137
Te Wheoro, Wiremu 9
Temm, Paul 35, **38**, **70**
Tepania-Ashton, Teresa **180–181**
Titokowaru, Riwha 32
Tomoana, Ngahiwi **172**, 184
training and development programmes 73, 83, 135, 141, **142**, 151
Treaty of Waitangi
    breaches 4, 6, 11, 153, 159–162
    guarantees 17, 19–20, 21–22, 25, 37
    interpretations 13–14, 21–26, 41–51, 113, 120
    Māori and English versions 22, 199–200
    promises 1, 6, 13–14

    recognition 27, 35, 71, 76, 79
    signing 1, **2–3**, 5–6, **15**
Treaty of Waitangi Act 1975: 35
Treaty of Waitangi (Fisheries Claims) Settlement Act 1992: 123, 141, 171, 197
Treaty of Waitangi Fisheries Commission *see* Te Ohu Kaimoana
Treaty Tribes Coalition **132–133**, 133–137
Tuhuru hapū 115, 118
Tuiono, Teanau **139**
Turia, Tariana 162, 165, **169**
Tuuta, Evelyn 124

U
urban Māori claims 135–138, **136**
urbanisation 10, 27

W
Waaka, Toro **128–129**, 141, 179
Waiariki District Māori Council 71
Waikato war 6
Waipapakura case 48–49
Wairau clash 6
Waitangi Tribunal 4, 11
    court rulings on scope 135–137
    development in 1980s 37–39
    establishment 27–35
    hearings **38**, 63–66, **64–65**, **70**, 164
    interpretation of Treaty 22–23, 25–26, 141
    report on aquaculture and marine farming 168
    report on foreshore and seabed rights 159–162, **164**
    report on Sealord deal 120
    reports on fishing rights 13–20, 22–23, 26, 35–39, **36**, 41–51, 57, 60, 63, 67–71, 88, 123
Wakefield, Edward Gibbon 5
Walker, Ranginui 27
Wallace, Justice J. H. 67, 71
Walzl, Tony 48
Wetere, Koro 63, 79, 123, **128–129**, 141, **178**, 179, **180–181**, 184
Whangarei County Council 32
Wharekauri Rēkohu 118
Whineray, Wilson 112
White, Greg **132–133**
Wickliffe, Caren 135
Williams, Joe 99
Williamson, Justice Neil 60
Wilson, Margaret **161**
Wilson, W. M. (Bill) 63
Wilson Neill 83
Wineti, Francene **142**
Winiata, Whatarangi 106, 112
Wyllie, Tukekawa **132–133**, 139
Wynyard, Manu 113